1906830

6/nn

15⁰⁰

TRACE ELEMENTS IN THE ATMOSPHERE

By H. Israël and G. W. Israël

Translated by
STS, Incorporated
Ann Arbor, Michigan

ann arbor science PUBLISHERS INC.
POST OFFICE BOX 1425 • ANN ARBOR, MICHIGAN 48106

To My Wife.

—Gerhard Israël

Foreword

The publication of this book falls into a time in which air pollution has assumed intolerable proportions in some places as a result of increasing population density, traffic and industrialization. Keeping the air clean has become one of the principal problems of the second half of the twentieth century.

The concept of atmospheric trace elements refers to all solid, liquid and gaseous contaminants of the atmosphere, which naturally also include anthropogenic air pollutants. A knowledge of the "life cycles" of these trace gases and aerosols represents a basis for an understanding of their possible effects on our environment.

It is the object of this volume to provide the reader with this basic information in a brief and generally comprehensible form. Following the characteristic of the series of "books of the Zeitschrift Naturwissenschaftliche Rundschau," it is directed simultaneously to the specialist and the layman in order to give both an overall view of our present knowledge and position in this field. The specialist is therefore asked to understand the necessary omission of numerous details in the discussions. The layman, in turn, should understand that it was not possible to omit all theoretical principles. However, these sections can generally be omitted without great detriment to an understanding of the following text.

The present volume represents the final result of the life work of my beloved father, which includes about 300 scientific publications and books. Unfortunately, he was not given the opportunity to complete this volume. I therefore was pleased to follow the last wish of my father and write the remaining chapters as well as to realize publication of the manuscript. My gratitude is due to my dear wife for her help in completion of the manuscript and reading of the proofs.

G. W. Israël
College Park, Maryland
1973

Table of Contents

Introduction

We customarily characterize the air surrounding us by certain data that express how its specific properties affect our sensory organs. First of all, we record it on the basis of its temperature and moisture content such as warm, cold, dry, and humid, thus giving information on its physical and thermodynamic state. At the same time, we react to the chemical state of the air and express this by information concerning its purity, using designations such as urban air, country air, sea air, and mountain air, which contain subjective general data concerning certain air pollutants that produce physical or psychological reactions. Thus, we distinguish the concept of *air* as a gas mixture of nitrogen and oxygen in a ratio of about 4:1 from the concept of *atmospheric air* with all of its additional characteristics.

Pollutants in the form of certain *trace* elements are the carriers of these various properties of air. This term summarizes all admixtures of impurities present in the gas mixture of nitrogen, oxygen and argon that is defined as air. Quantitatively, these impurities are present only in *very* small amounts, the gaseous components representing less than 0.1% of the total volume; only in extreme exceptions do aerosols attain a total mass of 10^{-3} g/m^3 of air.

As a rule, the sources of trace elements are found on the surface of the earth. From there, they reach the atmosphere and its systems of motion, characterized by wind and turbulent exchange, which distribute them in horizontal and vertical directions. They originate from soil erosion of rock material and sea spray which become airborne by atmospheric turbulence, smoke produced in natural fires, emission of volcanic gases and entrainment of biological products by the wind.

Artificial sources, related to the development of human civilization, are added to these natural sources. In many cases, they involve the same types of gases and aerosols as those produced by natural sources, but today they have in some cases significantly exceeded their abundance.

With regard to the nature of anthropogenic sources of materials introduced in the atmosphere, characteristic changes have taken place with time; as long as the need for energy generation was limited to the combustion of wood and other materials from the surface of the earth, this resulted only in a certain increase of materials originating from natural fires. A basic change occurred with the beginning of fossil fuel consumption (coal and later oil) and with increasing industrialization, which introduced impurities into the atmosphere that did not exist previously. Most recently, artificial radioactivity has been added as a new source.

The abundance of artificial sources, as mentioned, has in many cases exceeded that of natural sources. The familiar term *air pollution* is connected with their activity. Their continuous increase, particularly in our present industrial era, has assumed a proportion that has made them one of the major contemporary problems of civilization.

It is interesting that the first measure to control the introduction of "undesirable pollutants" in the air dates back to the thirteenth century with the beginning of coal heating in England. Legislative measures were taken against "the dangerous use of coal," and in 1306 even the death penalty was instated. In other countries, voices of concern about coal burning were also raised but this did not prevent its widespread use.

Only the tremendous increase of general air pollution of civilization in our time together with spectacular catastrophes with numerous deaths that occurred in this connection, for example, in the early thirties in the Meuse Valley or in 1952 in London, provided the incentive to devote systematic attention to this problem. Today, this subject is under careful and detailed observation and analysis in close collaboration between research and industry, the object being the reduction of air pollution or at least an arresting of its further increase.

In view of the significance of trace elements in the air, we shall report on facts that are known today. Natural and artificial sources will be treated separately, to the extent to which this is still possible.

The subject matter is organized so that gaseous pollutants are discussed first, in terms of their nature, origin, distribution and redeposition. The second part then deals with solid and liquid aerosols, their origin, growth, the microchemical reactions involved with them and their cycles in the atmosphere. In the third part, the group of radioactive materials in the atmosphere of natural and artificial origin is discussed.

Normally, trace gases and aerosols are distinguished as trace elements. However, this is not always possible exactly, since chemical reactions may be expected and do occur between the two as a function of existing conditions, favored primarily by the continuous presence of liquid and gaseous water in the atmosphere. The cycles between source, transport and deposition of the individual components consequently are frequently closely interrelated.

A distinction between trace gases, aerosols and radioactivity, on the basis of the subject matter to be discussed, requires a separation of presentation in parts of different volume, since investigations of trace gas distribution have become more extensive only in recent years in connection with gaseous air pollution, whereas considerably older and more extensive information exists on aerosol research.

Atmospheric Gases and Trace Gases

If all existing information on the gaseous composition of the atmospheric air is summarized, the survey shown in Table I is obtained. Table I applies to dry air near ground level. The individual gas components are expressed in ppm by volume (parts per million) corresponding to the amount of the respective gas in cm^3 per m^3 of air under normal conditions at 0°C and 760 mm pressure and in weight units of μg (10^{-6} g) per m^3. Water vapor is not included in the table since it holds a special position because of its phase changes in the atmospheric temperature range and cannot actually be considered a trace gas.

Table I indicates, then, that the three principal gases—nitrogen, oxygen and argon—represent 99.96% of atmospheric air and are always present in the same ratio of 78.09:20.94:0.93. They are joined by rare gases that are also always present in the same concentration, representing 0.024% of the total volume. The balance of less than 0.02% represents the large number of gaseous compounds appearing in variable concentrations. These are listed with their mean maximum and minimal values.

With regard to variable trace gases, the table is incomplete since it contains only those found always and everywhere. Gases that occur only occasionally in the closer vicinity of certain emission sources have been omitted. They will be discussed separately elsewhere in connection with the corresponding emitting sources, for example, motor vehicle traffic.

The fact that the concentrations of gases listed in the third group vary in place and time indicates that their presence is controlled by processes of differing effectiveness. It seems appropriate to interpret the respective concentration of a given trace gas as the state of equilibrium between its formation and destruction, or as an equilibrium between its introduction into the atmosphere and its subsequent removal from it. The data of the last column of Table I concerning the *mean residence time* of individual trace gases must be read in this context. Their definition results from the following considerations: If a certain reservoir, for example, the atmosphere, contains a total quantity M of a certain material

Table I

Summary of the Gas Composition of Dry (Water Vapor-Free) Atmospheric Air in the Troposphere*

Type of Gas	Chemical Symbol	Concentration		Residence Time
		ppm	$\mu g/m^3$	
Principal gases				
Nitrogen	N_2	780000	$976 \cdot 10^6$	continuous
Oxygen	O_2	209400	$298 \cdot 10^6$	continuous
Argon	Ar	9300	$16.6 \cdot 10^6$	continuous
Trace gases (constant)				
Helium	He	5.2	920	about $2 \cdot 10^6$ years
Neon	Ne	18	$1.6 \cdot 10^4$	continuous
Krypton	Kr	1.1	4100	continuous
Xenon	Xe	0.086	500	continuous
Trace gases (variable)				
Carbon dioxide	CO_2	200-400	$(4-8) \cdot 10^5$	4 years
Carbon monoxide	CO	0.01-0.2	10-200	about 0.3 yrs.
Methane	CH_4	1.2-1.5	850-1100	about 100 yrs.
Formaldehyde	CH_2O	0-0.1	0-16	?
Nitrogen oxides	N_2O	0.25-0.6	500-1200	about 4 years
	NO_2	$(1-4.5) \cdot 10^{-3}$	2-8	a few days
Ammonia	NH_3	0.002-0.02	2-20	?
Hydrogen sulfide	H_2S	$(2-20) \cdot 10^{-3}$	3-30	about 40 days
Sulfur dioxide	SO_2	0-0.02	0-50	about 5 days
Chlorine	Cl_2	$(3-15) \cdot 10^{-4}$	1-5	a few days
Iodine	J_2	$(0.4-4) \cdot 10^{-5}$	0.05-0.5	?
Hydrogen fluoride	HF	$(0.8-18) \cdot 10^{-3}$	0.7-16	?
Hydrogen	H_2	0.4-1.0	36-90	?
Ozone	O_3	0-0.05	0-100	about 60 days

*Table according to C. E. Junge, 1963.

and the quantity R is removed from it per unit of time, the following relation is valid at equilibrium:

$$\tau = M/R \tag{1}$$

where τ represents the mean residence time of the respective material in the reservoir under consideration. When applied to trace gases and aerosols, this value τ offers interesting information concerning the migration, distribution and life cycle of trace elements, as a few examples will show.

If trace gases are classified according to their residence times in the atmosphere, two groups become apparent: one part of the gases has residence times ranging from a few days to several weeks, while the others exhibit values of several years and more. Attempts have been made to correlate this grouping with the course of the water vapor cycle in the atmosphere, which is characterized by mean residence times of 1-2 weeks. Consequently, those trace gases that are redeposited with the reactive participation of atmospheric H_2O should have a cycle between formation and redeposition similar to the H_2O cycle. In contrast, it may then be assumed that the life cycle of gases with much longer residence times has little or no connection with the atmospheric water cycle.

Therefore a brief examination of the water vapor cycle in the atmosphere is presented before turning to a specific discussion of trace gas cycles.

WATER CYCLE IN THE ATMOSPHERE

Atmospheric air always contains a certain amount of water vapor. In the lower layers this varies between 0.004 and 4 vol.% (corresponding to 40 and 40,000 ppm or 3-3000 $\mu g/m^3$) as a function of climate and weather conditions. As we know, it drops rapidly with altitude.

The source of atmospheric water vapor is evaporation, while its sink is removal by precipitation after conversion into the liquid and solid phase. On the basis of a comparison of the total amount of water vapor present in a vertical column of air—so-called precipitable water—and the total amount of precipitation deposited on the surface of the earth in a certain period of time, Equation 1 leads to a mean residence time τ of about 10 days with a marked dependence on latitude (see Table II).

The numerical differences demonstrate the latitude-related differences in climate: the relatively low precipitations of the tropical latitudes and polar regions are clearly distinguished by their slower H_2O cycle as compared to the tropics and particularly the high-precipitation west-wind zones of the middle latitudes.

Table II

Mean Residence Time of Water Vapor
in the Atmosphere as a Function of Geographical Latitude*

Latitude	Mean Residence Time of Water Vapor in Days
0-10	8.1
10-20	11.2
20-30	12.0
30-40	8.7
40-50	6.4
50-60	6.2
60-70	8.7
70-80	(13.7)
80-90	(15.0)

*According to C. E. Junge, 1963.

TRACE GASES AND THEIR CYCLES

According to Table I, the variable content of trace gases that are always present in the atmosphere consists of gaseous carbon, nitrogen and sulfur compounds, halogens, ozone and hydrogen. It may be assumed that each of these gases has a certain life cycle that leads from its formation and introduction into the atmosphere to its disappearance from it as a result of chemical reactions and redeposition. During its residence in the atmosphere, the respective gas is often already modified by chemical and photochemical reactions, and these processes may be linked with aerosol cycles.

The following discussion will present an analysis of these cycles in detail, to the extent to which these are apparent at this time.

The CO_2-cycle

Among the carbon compounds CO_2, CO, CH_4 and CH_2O listed in Table I, carbon dioxide holds the predominant position. With its concentration of 200-400 ppm, it at the same time represents the major one of atmospheric trace components in general.

In terms of its formation and degradation, atmospheric CO_2 is directly related with the biosphere. Its sources and sinks are located accordingly on the surface of the earth, practically exclusively in the continental regions. Although the ocean is the main reservoir of terrestrial CO_2, practically it represents only an inactive buffer zone.

Over continental regions, CO_2 is formed during the decomposition of organic matter: all processes of combustion, decomposition and degradation of such matter release CO_2 into the atmosphere. The CO_2 supplied by volcanic exhalation and mineral sources as well as gas release from lower geological layers plays only a subordinate role. CO_2 degradation takes place in the assimilation process of the plant world.

The global turnover of this cycle can be estimated to amount to an annual consumption of 3% of the mean atmospheric carbon dioxide content for incorporation of carbon into the biosphere. According to Equation 1, therefore, the residence time relative to degradation is characterized by a value of $\tau_{ab} = 33a$ (see Figure 1).

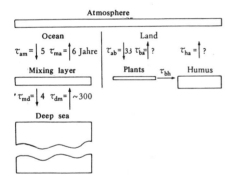

Figure 1. Schematic diagram of the carbon dioxide cycle (according to H. Craig, 1957 and C. E. Junge, 1963, see text for details).
Indices:
a: atmosphere
b: biosphere (plant world)
h: humus
m: mixing layer in ocean
d: deep sea

The oceans form the main reservoir of terrestrial CO_2, with CO_2 exchange between the two reservoirs taking place on the surface of the water. Since the CO_2 concentration of sea water is influenced by the temperature, salt content and pH of the water, it may be assumed, in accordance with these relationships, that the cold masses of water of the high latitudes act primarily as sinks, while the warm waters of the tropics and lower latitudes primarily represent sources of atmospheric CO_2. The balance is produced by deep-water currents in the ocean, in which the cold polar water masses move under the climatically warmer zones and thus transport water of higher CO_2 content into the lower latitudes. The biospheric cycle of the oceans need not be considered here, since it is essentially closed in itself.

In the exchange process between atmosphere and ocean, a specific distinction must be made between an upper mixing layer of the oceans of about 50-100 m depth and the actual deep sea.

If our knowledge to date concerning CO_2 movements in the atmosphere, ocean and biosphere is summarized, the entire CO_2 cycle can be described schematically approximately as shown in Figure 1 according to H. Craig, 1957 (see also C. E. Junge, 1963).

The right side of Figure 1 shows the active sources (*i.e.,* producing or degrading CO_2) and sinks of CO_2 presumably present in the continental biosphere, while the left side represents the effect of the marine reservoirs participating in the cycle with their biologically inactive sources and sinks. The atmosphere is the mediator between the two zones.

The course of the phases of the cycle with time is characterized by individual τ-values. The sequence of the two indices shows the direction of the CO_2 motion: τ_{am} thus refers to the kinetic process of CO_2 transport from the atmosphere, a, into the oceanic mixing layer, m. The height of the rectangles shows the mean carbon dioxide concentration M_x of the respective reservoir x expressed in grams for a column of 1 cm^2 cross section. The respective numerical values are as follows:

$$M_a = 0.48 \text{ g per } cm^2$$
$$M_b = 0.22 \text{ g per } cm^2$$
$$M_h = 0.81 \text{ g per } cm^2$$
$$M_m = 0.55 \text{ g per } cm^2$$
$$M_d = 26.8 \text{ g per } cm^2$$

Radioactive carbon, ^{14}C, plays a decisive role in the determination of individual time constants, particularly the course characterizing the atmospheric-oceanic exchange process. In the atmosphere, ^{14}C is found by cosmic rays. After oxidation to $^{14}CO_2$, it labels normal atmospheric CO_2. It reaches the plant world with the latter by assimilation and from there the human and animal body with food.[1] In the course of time, this process has resulted

---

[1] Carbon-14 is formed by thermal neutrons (n) produced by cosmic rays, reacting with nitrogen atoms with release of a proton (p) according to the reaction:
$$N^{14} + n = C^{14} + p$$
and decays into ^{14}N with β-emission (e⁻) with a half-life of 5760 years:
$$C^{14} \ldots \ldots N^{14} + e^-$$
The forming ^{14}C is oxidized relatively rapidly into
$$C^{14}O_2$$
in the atmosphere. It mixes with atmospheric CO_2 in this form.

in labeling of the entire carbon of the organic world with a certain fraction of ^{14}C. According to the studies of H. E. Suess (1955) with wood from the 19th century, the ratio of ^{14}C to normal ^{12}C has the following value:

$$^{14}C:{}^{12}C = 1.24 \cdot 10^{-12}:1$$

The age of a given object can be determined b, the changes of this ratio in certain carbon samples by the well-known carbon dating method. In the present case, studies of the $^{14}C/^{12}C$ ratio in oceanic CO_2 with consideration of certain additional effects, such as isotope separation effects and Suess-effect, furnish information concerning the time constants τ_{am}, τ_{ma}, τ_{md} and τ_{md} (see Figure 1).

The M_x-values listed above in connection with Figure 1 are valid under the assumption that a steady state is reached between the individual phases of the cycle. Short-term variations such as those caused by the diurnal-nocturnal rhythm, seasonal fluctuations and meteorological processes, can be neglected. Moreover, in view of the value of the time constants that are counted in years in the various exchange zones of the total cycle, the slow horizontal exchange from the Northern Hemisphere of the atmosphere across the equator also does not modify the picture since it is still relatively rapid.

The situation is entirely different for slow (secular) variations of the CO_2 concentration. Thus, the increase of atmospheric CO_2 concentration that has been observed for about 100 years as a result of increasing industrialization represents a disturbance of the equilibrium state. Figure 2 shows this increase since 1870.

The rise of about 14% is in good agreement with the increase in fuel consumption in the same period (R. Revelle and H. E. Suess,

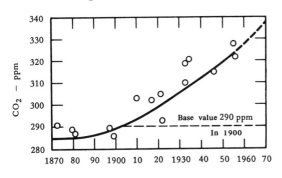

Figure 2. Increase of the mean atmospheric CO_2 concentration in the period from 1870 to 1956 (according to G. S. Callendar, 1958).

1957). If we assume that the natural sources and sinks have remained unchanged, this parallel allows us to estimate the anthropogenic CO_2 production.

Since the production of CO_2 by civilization is the result of the use of fossil fuels, *i.e.*, ^{14}C-free, a decrease of the $^{14}C/^{12}C$ ratio in the atmosphere may be expected. This is actually observed in the form of the so-called Suess effect, but with an approximately 2-3% decrease of the ratio, it is notably smaller than the general mean CO_2 increase. The discrepancy between these two values has not yet been explained to satisfaction.

Analytical methods are made difficult today due to the atmospheric nuclear weapons tests, which modified the $^{14}C/^{12}C$ ratio. Figure 3 shows the ^{14}C-concentration of the atmosphere from 1400 to the most recent past. We can recognize the decrease that has occurred for about 100 years as a result of the Suess effect on one hand and the increase since the early fifties due to nuclear weapons tests on the other hand.

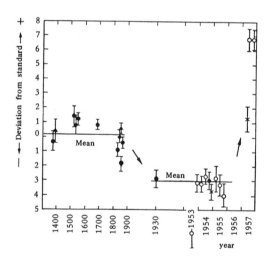

Figure 3. ^{14}C concentration of the atmosphere from 1400 up to the most recent past (1957) according to K. O. Münnich (1957) as well as K. O. Münnich and J. C. Vogel (1958). (Plotted in % deviation from the "standard value.")

When the above M_x-values are multiplied with the respective surface values, they furnish information concerning the total global CO_2 concentration. Thus, the total atmospheric CO_2 content is calculated to be $2300 \cdot 10^9$ t, while that of the oceans amounts to approximately 43 times this value.

A. C. Stern cited the following estimated values for the total turnover: annual consumption by assimilation approximately

$60 \cdot 10^9$ t; release by respiration and decomposition of organic matter, approximately the same; introduction by combustion processes approximately $9 \cdot 10^9$ t.

Other Gaseous Carbon Compounds

Our knowledge concerning other gaseous carbon compounds in the atmosphere is considerably more limited.

Carbon monoxide (CO) also belongs among the continuously present trace gases. Since it is subject to considerably more rapid variations with time compared to CO_2, its cycle must be characterized by smaller time constants.

The sources of CO are located at ground level. The major contribution presumably is made by anthropogenic combustion processes and about 85% of these derive from the automobile. This is clearly apparent from the following data: concentrations of 100 ppm have been measured on heavily traveled roads, an average of 5 ppm in cities and 0.08 ppm in open country. In the oceans, carbon monoxide is presumably formed by the decomposition of organic matter, with some of it escaping to the atmosphere. The oceanic CO contribution appears to be in the same order of magnitude as the anthropogenic imission (Seiler, Junge, 1970).[2] The mean residence time of carbon monoxide in the atmosphere is estimated to be 100-300 days. In addition, it is believed that an annual increase of the total CO concentration of about 0.03 ppm is taking place under present-day conditions.

Little definite information is available concerning the decomposition and redeposition of CO. Its oxidation into CO_2 in the presence of atomic oxygen takes place practically only at altitudes of the ozone layer and above. Certain soil bacteria also seems to be capable of inducing this oxidation. It is therefore assumed that the stratosphere and the soil represent the major sinks of carbon monoxide.

Methane (CH_3) is found in relatively high concentrations. Its presumed sources are natural formation during the decomposition process of animal matter in swamps and lakes under the influence of anaerobic bacteria as well as natural gas, while artificial sources are present in wastewater and sewage. The annual total imission is

---

EDITOR'S NOTE: In this publication, "emission" refers to emitted substances in the atmosphere, and "imission" to emitted substances which have settled to a surface. In U.S. literature, "emission" is commonly used for both terms.

estimated to be 10^7-10^8 t (G. E. Hutchinson, 1954). Together with the total atmospheric concentration, this results in a lifetime τ of about 100 years, which is also confirmed by ^{14}C dating.

The decomposition of CH_4 takes place at higher altitudes of the atmosphere. Reactions with atmospheric ozone and atomic oxygen in the stratosphere and mesosphere may be considered certain but not sufficient in their total effect. It is most probable that CH_4 is decomposed by radiation with wavelengths shorter than 1450 Å, which may penetrate the atmosphere from outside up to altitudes of about 70 km. Methane-degrading processes at ground level are not known. In particular, CH_4 is "immune" to bacterial action.

Formaldehyde (CH_2O) is found in very low concentrations. Little is known about its origin. It may form in incomplete combustion processes and under the influence of solar radiation on aqueous organic solutions as well as by photochemical dissociation of CO_2 and CH_4 at altitudes above 70 km. Its photolysis by irradiation with light of less than 3600 Å wavelength, *i.e.,* in the atmosphere down to ground level, is possible.

Other hydrocarbon compounds that are occasionally found, such as ethylene, propane, butane, terpenes, will be discussed later.

Sulfur Gases

The conditions of sulfur gases are more complicated because of their much higher chemical activity compared to the carbon compounds discussed above. In the atmosphere, sulfur is present in several compounds, namely, in the form of sulfur dioxide (SO_2) and hydrogen sulfide (H_2S), both gases, and sulfur trioxide (SO_3), the oxidation product, as well as sulfuric acid (H_2SO_4), formed from the latter by water absorption, and its salts. In the latter form it becomes an important component of atmospheric aerosols. Proportionately, an average of 80-90% of the atmospheric sulfur is present in the two gaseous compounds. This amount may drop to 70% and less only in a purely maritime atmosphere, for example, in Hawaii (H. W. Georgii, 1960).

The sources of atmospheric sulfur are located on the surface of the earth. Some are natural and others are artificial. From the standpoint of air pollution, SO_2 production of highly populated areas and industrial centers naturally predominates as is clearly evident in the graph of H. W. Georgii (1960) shown in Figure 4.

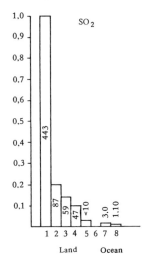

Land   Ocean

Figure 4. SO$_2$ concentrations at various stations according to H. W. Georgii (1960). Height of individual bars in proportional relative units. The numerical data are expressed in $\mu g/m^3$ of air.
1. Frankfurt/Main–Inner City in winter (November-March)
2. Same monitoring station in summer (April-October)
3. Taunus–Observatory (winter)
4. Zugspitze (August)
5. St. Moritz (August)
6. Round Hill
7. Florida
8. Hawaii

In spite of this marked preponderance of the civilization-related fraction of atmospheric SO$_2$, it has basically only local significance. Although the SO$_2$ concentration may increase locally up to a physiological nuisance value (threshold at about 1000 $\mu g/m^3$), it rapidly decays with distance from the source by dilution and relatively rapid redeposition. The civilization-related contribution represents a notable but by no means preponderant partial fraction of the total atmospheric sulfur balance as indicated by Figure 5.

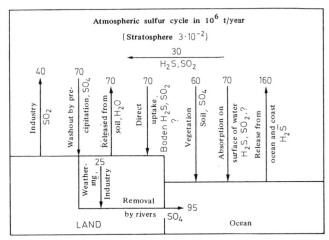

Figure 5. Sulfur cycle in global representation according to C. E. Junge (1963).

The natural sources of atmospheric sulfur consist primarily of the decomposition of organic matter on and in the soil, apart from a small fraction in volcanic exhalations. Decomposition takes place primarily in the absence of air (or oxygen) in the soil, on shallows, in shelf floors near the ocean coastlines fed by river water, and in the oceans themselves. The H_2S formed in this process first reaches the atmosphere as such and is relatively rapidly converted into SO_2 and higher oxidation states. It may be assumed that the ozone concentration of the atmosphere is of particular importance in the process.

The sinks of atmospheric sulfur are also located on the surface of the earth. Redeposition takes place essentially by precipitation, which takes up sulfur partly by dissolution of gaseous SO_2 and H_2S, but for the most part probably after oxidation into sulfate, and returns it to the soil or ocean.

Although numerous questions remain concerning these partial processes, which cannot be discussed in detail here, the sulfur cycle can probably be described qualitatively and quantitatively by the diagram of Figure 5.

H. W. Georgii (1967) offered the representation shown in Figure 6 of the presumed altitude distribution of SO_2 over land and sea. The two peaks over land near the surface characterize the conditions for winter (Wi) and summer (So).

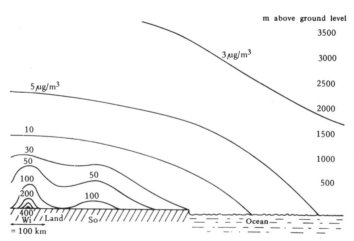

Figure 6.   Schematic representation of the SO_2 concentration over land and sea according to H. W. Georgii (1967).

Nitrogen Compounds

Gaseous nitrogen compounds are found in the atmosphere in the form of nitrogen oxides, N_2O (nitrous oxide, "laughing gas"), NO (nitrogen monoxide) and NO_2 (nitrogen dioxide, nitric anhydride), as well as in the compound NH_3 (ammonia).

It is highly probable that N_2O forms mainly in the soil by the breakdown of nitrogen-containing organic matter under the influence of bacteria, with the conversion of NH_4^+ and NO_3^- ions into N_2O and N_2 (see, for example, P. W. Arnold, 1954). The total production of N_2O has been estimated to be about $1.6\text{-}16 \cdot 10^{10}$ molecules per cm^2 and sec ($1.2\text{-}12 \cdot 10^{-12}$ g/cm^2 sec) by R. M. Goody and C. D. Walshaw (1953).

Since N_2O is a very stable and chemically inert gas, it undergoes no chemical conversions in the atmosphere. Its destruction takes place by photodissociation at wavelengths shorter than 2100 Å, *i.e.*, it may be expected only at altitudes above the ozone layer. Its mean residence time in the atmosphere has been estimated at four years.

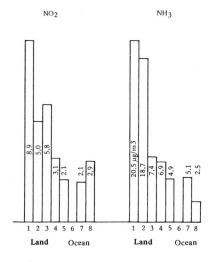

Figure 7. NO_2 and NH_3 concentrations at different monitoring stations according to H. W. Georgii (1960). (Details of presentation and monitoring stations are the same as in Figure 3.)

There are two probable sources of NO_2, of which one is related to population density as indicated by Figure 7 on the left. At the same time, however, we may assume a natural production by plant biology (as indicated by a notable NO_2 formation in silos of agricultural products), for which bacterial oxidation of NO into NO_2 may be assumed. On the basis of the nitrate content of precipitation, which suggests a certain correlation with electrical processes in the atmosphere, direct NO_2 formation by electrical discharges also seems possible (R. and M. Reiter, 1958).

The redeposition of NO_2 probably takes place by condensation and precipitation

processes after it has been converted into nitrate. Its residence
time in the atmosphere is estimated at two months (C. E. Junge,
1963).

NO is observed in some places of high air pollution, particularly
in Los Angeles where up to 3 ppm ($NO + NO_2$) were occasionally
detected in the winter of 1960-61 (A. C. Stern, 1968).

Ammonia appears in gas form (NH_3) and in dissolved form as
NH_4^+ ion in precipitation. As in the case of SO_2 and NO_2, the
gaseous form as a rule predominates over the hydrolyzed fraction.
The sources of atmospheric NH_3 are also found on the surface of
the earth and result partly from natural processes and partly from
civilization. Depending on its pH and water content, the soil re-
leases varying amounts of NH_3. It is formed in the soil from bio-
logically-bound nitrogen and amino acids under bacterial action.
Thus far, the involvement of the oceans in the atmospheric NH_3
balance has not yet been fully explained. It is possible that sources
and sinks are involved. The latter is suggested by the decrease of
the NH_4^+ concentration in the upper 100 m of water, which indicates the presence of an absorption process. However, this decrease
may also be attributable to NO_3^- reduction by planktons, the
activity of which must decrease with depth due to the loss of
light. Furthermore, films of organic matter on the water surface
may participate significantly in the NH_3 exchange between atmosphere and ocean.

The influence of population density is shown on the right of
Figure 7. In the metropolis of Frankfurt, the winter and summer
values are nearly identical. This indicates that NH_3 production is
the result of certain industrial waste gases rather than of heating.

The main sink of atmospheric NH_3 is probably its transport
back to the surface of the earth by precipitation.

Halogens

Among the halogens, chlorine, iodine and fluorine have been
detected in the atmosphere. According to the measurements of
H. W. Georgii (1960), a continental (anthropogenic) and a maritime source can be clearly recognized for chlorine (see Figure 8).

A decision as to whether chlorine is present or predominates in
the form of chlorine gas (Cl_2) or chlorine ion (Cl^-) or hydrogen
chloride (HCl) is difficult. The most probable nonanthropogenic
source of atmospheric chlorine appears to be the conversion of
NaCl from salt-water spray with sulfuric acid in cloud droplets or
aerosol particles according to the reaction (E. Eriksson, 1959):

$$2\,NaCl + H_2SO_4 = Na_2SO_4 + 2\,HCl \qquad (2)$$

No details are known concerning anthropogenic production.

Iodine is probably present as I_2 and some of it may reach the atmosphere due to emissions of the iodine-processing industry. According to B. Bolin (1959), the ocean must also be viewed as a source of atmospheric iodine, and an ocean-atmosphere-biosphere cycle is probable.

Fluorine in the form of HF is occasionally found in highly polluted urban air as well as in rural regions (A. Stern, 1968). Combustion of fluorine-containing fuels and certain industrial waste gases are considered to be its source.

The redeposition of halogens occurs in the form of precipitation and, in the case of iodine, perhaps also by direct uptake by soil and plants.

Other Gases

Table I contains three other trace gases for which formation and redeposition or decomposition remain in equilibrium: ozone (O_3), hydrogen (H_2) and helium (He).

Ozone (O_3) forms in the stratosphere by photochemical processes: short-wave sunlight with wavelengths of less than 2400 Å

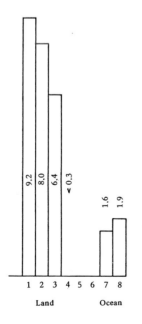

Figure 8. Cl_2 concentrations at different monitoring stations according to H. W. Georgii (1960). (Details of presentation and monitoring stations as in Figure 3.)

9.2 8.0 6.4 ◄ 0.3

1.6 1.9

1 2 3 4 5 6 7 8

Land   Ocean

can dissociate the oxygen molecules according to the reaction:

$$O_2 + h\nu \rightarrow O + O \tag{3}$$

The collision of an O_2-molecule and an O-atom in a so-called three-body collision, *i.e.*, with the participation of any other molecule or atom M, leads to formation of O_3 according to the reaction:

$$O_2 + O + M \rightarrow O_3 + M \tag{4}$$

This process of formation is counteracted by annihilation processes:

$$\begin{aligned}
O_3 + O &\rightarrow 2\,O_2 \\
O_3 + O_3 &\rightarrow 3\,O_2 \\
O_3 + O_2 &\rightarrow O + 2\,O_2
\end{aligned} \tag{5}$$

and by Process (6), which takes place under the influence of radiation at wavelengths shorter than 1100 Å:

$$O_3 + h\nu \rightarrow O_2 + O \tag{6}$$

In the high atmosphere, Reaction (3) finally leads to complete dissociation of molecular oxygen. Because of the need for the three-body collision, Reaction (4) can occur only at higher pressure, *i.e.*, at lower altitudes.

If we now calculate the ozone concentrations to be expected in the atmosphere as a function of altitude with consideration of the reaction probabilities and their dependence on pressure and temperature as well as the absorption coefficients in these reactions, we arrive at the result of Figure 9 (dashed curve).

The position of the maximum and the decrease in ozone concentration toward higher altitudes thus are well described by theory. However, in the direction of lower altitudes the differences between calculation and observation are considerable.

The findings concerning total ozone concentration and its variation suggest relationships between the troposphere and stratosphere: while we should anticipate a direct correlation with the altitude and position of the sun as well as its periodic diurnal and annual variations and a corresponding dependence on latitude, measurements reveal considerable short-term, long-term and

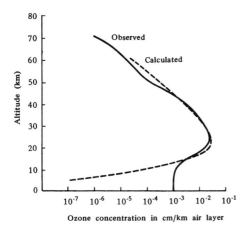

Figure 9. Altitude distribution of the atmospheric ozone concentration according to calculation (dashes) and measurement (solid curve).[3]

meridional differences which can *not* be explained by changes of the photochemical reactions.

The short-term diurnal variations suggest a relationship with meteorological conditions in the troposphere. Although measurements are not yet sufficient to offer a clear picture of the mechanism of this interrelationship, a rule can be formulated stating that a cold troposphere, a low tropopause and a warm lower stratosphere are related with a high ozone concentration and vice versa. This leads to the conclusion that air masses of different origin evidently "carry along" their properties into the ozone region even when they move over longer distances. The meridional distribution of the total atmospheric ozone concentration and its seasonal variability are shown in Figure 10.

The calculated and measured altitude distribution of ozone below the maximum region in the stratosphere shows a discrepancy in Figure 9 because in this zone ozone transport evidently has a downward direction. This is the result of atmospheric vertical exchange processes that extend up to this altitude and transport a significant fraction of stratospheric ozone from the stratosphere

---

[3] By convention, the atmospheric ozone concentration is expressed in cm per total atmosphere or (with topographical contour lines) in cm per km. This value means that the respective quantity of ozone in an air column of atmospheric altitude or of 1 km altitude, converted to normal conditions of $0°C$ and 760 mm pressure, would have a certain height in cm. For the conversion, it may be considered that a quantity of ozone of 10^{-3} cm per km expressed in this form corresponds to a concentration of 21.4 μg per m^3 or about 0.01 ppm.

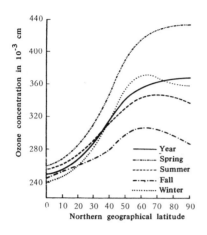

Figure 10. Meridional distribution and seasonal fluctuation of the mean ozone concentration [according to J. London (1962)].

into the troposphere. On contact with oxidizable material in the clouds, tropospheric trace elements (for example, the above-mentioned reactions with CO, CH_4 and H_2S) and materials on the surface of the earth, this fraction is dissociated by chemical processes.

An attempt can be made to characterize this part of the ozone cycle by corresponding τ-values, resulting in the following approximate values (see, for example, C. E. Junge, 1961):

  a. lifetime of O_3 in the (lower) stratosphere of up to about two years
  b. lifetime of O_3 in the troposphere about 1-1.5 months
  c. ozone introduction into the troposphere from higher layer: total of about 0.1-0.2 μg per m^2 and sec.

Possible natural ozone sources are believed to be discharge processes of the atmospheric electrical field of high field strength. Furthermore, photochemical ozone formation is observed in urban air pollution, for example, in Los Angeles, Albuquerque, N.M. and elsewhere.

Hydrogen (H_2) in the atmosphere has various sources. First of all, it forms during decomposition of organic matter in the soil by anaerobic bacterial action. Second, in the high atmosphere, the H_2O molecule is subject to photodissociation at wavelengths of 1600-1800 Å. Finally, a possible source is represented by the introduction of solar protons.

Hydrogen sinks are represented by photo-oxidation in the presence of oxygen and ozone, bacterial utilization for protoplasm synthesis and reduction of nitrites and amino acids, and finally escape from the exosphere at altitudes above 500 km into space.

Precise data concerning these processes that might characterize the details of the hydrogen cycle are not yet available.

Helium (He) originates from radioactive α-decay in the decay series that begins with uranium-238, uranium-235 and thorium-232. Since the total atmospheric concentration of helium is considerably smaller than it should be if it had to accumulate since the origin of the earth, it may be assumed that it escapes into space in the exosphere.

Estimates of the total atmospheric helium concentration based on a comparison with its formation and the assumption of corresponding emanation from upper layers of the earth lead to a residence time of about $2 \cdot 10^6$ years as listed in Table I.

3

Atmospheric Aerosols

The solid and liquid trace elements present in the atmosphere as a rule are known by the general term "aerosols." This term originated as an analog to hydrosols and designates solid or liquid particulates that are suspended in a gaseous "solvent."[4] Other terms, such as "air colloids" and "air plankton" are less common but otherwise have the same meaning.

The concept of aerosols became generally accepted after publication of the monograph "The Atmosphere as a Colloid" by A. Schmauss and A. Wigand (1929). It is the first description of particles suspended in the atmosphere as a whole considered from a uniform point of view and dealing with their significance in meteorology.[5]

Aerosols have a universal significance in terrestrial history. First of all, they represent the starting points for condensation and sublimation of water vapor. It can be easily understood that condensation or sublimation of water vapor would be an entirely different process, if it were possible at all, in the absence of such suspended particles that are actually present in the real atmosphere. In any case, the water cycle as we know it would not exist. Consequently, all phenomena related with this cycle, such as weather and climatic conditions, radiation and thermal equalization, and weathering, would not occur or would be reduced or modified in such a way that biological life as we know it would hardly be possible. It may be assumed that the surface of the earth would look like the pictures of the moon and Mars.

---

[4] It is customary to call the entity of solvent and particles the hydrosol and aerosol, respectively, while the particles themselves are called the hydrosols and aerosols, respectively.

[5] Schmauss and Wigand introduce their study as follows: "Only rarely is the atmosphere a molecularly disperse system. It can claim this only after prolonged rain or snowfall. Usually, it is a colloid-disperse system with the character of gas as solvent, solids or liquids dissolved or suspended in it in finest distribution. The particle size will also have its significance in the atmosphere as in the case of hydrosols. In analogy with this expression, we will speak of an *aerosol* in the following."

By their presence, aerosols give atmospheric air its characteristic features: since they enter the atmosphere from the surface of the earth, they label moving air masses with particulates that reflect the immissions of different source regions that are traversed. Consequently, characteristic properties are imposed upon it that offer additional meterorological information for a classification of individual air masses, and their history. Because of the continuous change in air masses due to weather phenomena, the aerosol state becomes an important meteorological parameter distinguished by considerable variations and responsible for numerous separate phenomena: the visibility range as well as the intensity of the blue color of the sky depend on this state. Attenuation of solar radiation in its penetration through the atmosphere is closely related with the atmospheric aerosol content. Furthermore, the electrical properties of air are decisively controlled by the aerosol.

Aerosols are also of great importance for problems concerning atmospheric influences on biological processes, particularly human welfare, since they reach the organism by various pathways—through food and respiration. In this process they have immediate effects or may become physiologically active in the form of carriers of addition products (trace gases, radioactivity), depending on their physicochemical nature.

Finally, as a function of their accumulation and nature, aerosols can serve as a criterion of the degree of anthropogenic air pollution. For a long time, the small aerosols ("Aitken nuclei," see Table IV, page 32), which can be determined relatively easily, have been used as a criterion of air quality. It continues to be an overall criterion, even though pollution can be analyzed with considerably more differentiation, since all air pollution processes are accompanied by an increase of the imission of such small aerosols.

Atmospheric aerosols owe their existence to a large number of processes that result in mass exchange between the surface of the earth and its air envelope. A significant proportion of aerosols originates from volcanic activity and combustion processes of all types, among which the influences of population density, industry and motor vehicle traffic stand out. Moreover, mineral and organic matter is entrained by air motion and spray-water droplets rise from surf, wave crests and related processes. Furthermore, penetration of cosmic matter in dispersed form from outside has also been demonstrated. In addition, minute electrically charged particles form by ionization, followed by molecular cluster formation and the introduction of natural and artificial radioactive materials.

All of these finely dispersed materials are distributed horizontally and vertically in the atmosphere by wind and exchange. Accordingly, we find them *everywhere* in the atmosphere, over the oceans and in the polar regions, sometimes at a great distance from their sources.

As is true for trace gases, a certain life history can be followed in aerosols from their formation via their transport up to their redeposition. Numerous partial processes interact that form, move, modify and redeposit the aerosol. A schematic diagram of these processes is shown in Figure 11. The terrestrial sources are indicated at the left bottom and the cosmic source at the left top. The center portion shows the various transport possibilities of air motion and exchange as well as modifications by coagulation and chemical reactions with trace gases. On the right, the redeposition processes are sketched in which embedment in clouds and particles with subsequent rain-out command first place.

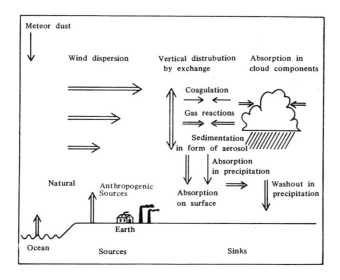

Figure 11. Schematic diagram of trace element cycles in the atmosphere (according to H. W. Georgii, 1965).

Figure 12 shows the cleaning process by clouds and precipitation. The smaller particles are incorporated in cloud droplets by

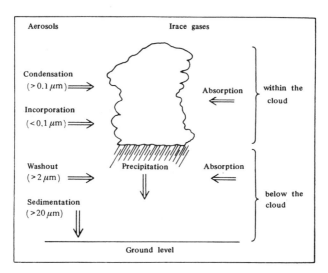

Figure 12. Schematic diagram of the cleaning process by clouds and precipitation for aerosols (left) and trace gases (right) according to H. W. Georgii, 1965.

condensation and addition and are returned to the surface of the earth by falling precipitation. The larger particles are washed out by falling precipitation or, with increasing size, are removed by sedimentation. (The right part of the figure also applies to trace gases.)

Viewed in terms of weight, the aerosol content of the air—as emphasized previously—is very small. Amounts of about 5-50 μg (micrograms) per m^3 are present in the form of natural aerosols. In populated regions with a polluted atmosphere, the values are 10-100 times higher with a clear dependence on population density: increase of about 50 to 100 $\mu g/m^3$ in areas with low population density, about 100-200 $\mu g/m^3$ in small and medium towns, up to 200-800 $\mu g/m^3$ in a metropolis and in industrial centers. (In London, up to 4000 $\mu g/m^3$ were detected under severe smog conditions.)

Table III contains data concerning the frequency distribution of the aerosol mass concentration in urban and rural air according to American studies. This distribution of the aerosol concentration clearly indicates a shift of the peak in the size spectrum in the direction of larger particles in the transition from rural to urban air;

Table III

Frequency Distribution of the Aerosol Content of Urban and Rural Air According to U.S. Studies*

Aerosol Content in $\mu g/m^3$	Urban Air	Rural Air
< 20	0.1%	22.0%
20- 40	2.0%	40.0%
40- 70	12.0%	28.0%
70-100	16.0%	7.0%
100-200	43.0%	2.8%
200-300	17.0%	0.2%
300-400	8.0%	
> 400	3.9%	

*Based on a tabulation of A. C. Stern, 1962.

this suggests that larger quantities of coarse aerosols evidently are also formed in urban pollution by smoke and combustion products.

In the following chapters, a differentiated description of aerosols will be given by a review of their processes of formation, size distribution, various physical and chemical properties and distribution in the atmosphere. Although a division of the description into separate discussions of natural and anthropogenic aerosols would be desirable just as for the trace gases, this is no longer feasible today. Instead, we can only distinguish between a "background aerosol," which is always present, and special local "air-polluting aerosols," which are modifications produced by local emissions.[6]

On the basis of this classification, the following sections dealing with the physics, chemistry, meteorology and climatology of aerosols will offer a summary of the properties of atmospheric background aerosols that are continuously present. The special air-polluting aerosols will then be treated separately in a later part of the text.

---

[6] A distinction can be made approximately in such a way that the aerosols present over the oceans and continents at more than 3 km altitude are considered the background aerosols, regardless of whether the latter are of natural or anthropogenic origin.

Natural and artificial radioactivity will be discussed in a separate chapter since it assumes a special position in several respects, even though it also belongs to the subject of atmospheric trace elements.

PHYSICS OF AEROSOLS

The particle size spectrum of aerosols extends from molecular dimensions up to about 100 μm, *i.e.,* it covers a range of more than 5 orders of magnitude. This rules out any standardized method of analysis, particularly because the lower range of this spectrum cannot be observed directly in the microscope. In principle, the following methods can be used for a size determination.

As long as the particles are microscopically visible, they can be directly observed as to size and shape (see, for example, Figure 13a). This is always the case for particle diameters of about 1 μm and more. The measuring range can be extended to smaller sizes by one order of magnitude by means of the ultramicroscope, which makes it possible to recognize the particles and to determine their size from the intensity of the scattered light produced by them (Figure 13b). However, no information is obtained about particle shape. The use of the electron microscope leads to the identification of dimensions that are at least one factor of ten smaller, but at the same time also allows a determination of particle shapes (Figure 13c). However, this technique is limited because some particles can be partially vaporized during electron bombardment or by the vacuum, so that they can be identified only by the tracks left on the substrate.

Figure 13a shows a partial view of an aerosol deposit of microscopically visible particles obtained by means of a so-called "conimeter." Figure 13b is a dark-field photograph of smog aerosols in Los Angeles with particle sizes of less than 1 μm (photograph of A. Goetz). Figure 13c is an electron micrograph of a continental aerosol (from C. E. Junge, 1963).

Thus, practically only indirect methods can be used for the identification and size determination of aerosols in the size range of less than 0.1 μm. Two very important methods are the *ion technique* and the *condensation technique.* As long as the particles carry an electrical charge, their number and size can be derived from the electrical current generated by them under the influence of an electrical field (ion technique). This technique can be

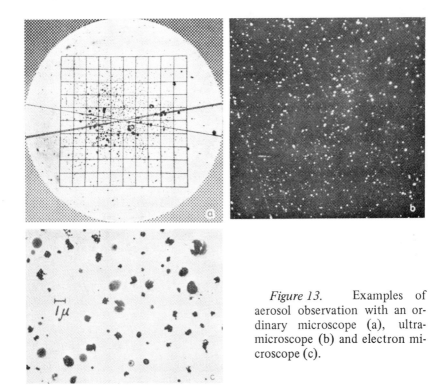

Figure 13. Examples of aerosol observation with an ordinary microscope (a), ultramicroscope (b) and electron microscope (c).

successfully used up to particle sizes of 0.01 μm, but it loses reliability for larger dimensions. In the condensation technique, supersaturated water vapor is generated in the vicinity of the particles. The water vapor condenses on the aerosols and they grow into droplets of visible size.

Diffusion and coagulation determinations can also furnish information. However, the results become less accurate with increasing heterogeneity of the aerosol size distribution.

It is important that these various methods of size determination overlap and can therefore be used in sequence. By combining the various techniques in one test system, it is possible to measure a broader range of the spectrum simultaneously (for example, see K. T. Whitby and W. E. Clark, 1966, measurements in the range of 0.015-1 μm).

In accordance with the different physical characteristics of aerosol particles that have resulted in the above-described methods for their determination, they are usually classified by size as shown in Table IV. The last column lists the sectors in which a given size class is of particular importance.

Table IV

Summary of Nomenclature and Special Significance of the Individual Ranges of the Aerosol Size Spectrum

Size Range in μm	Name	Special Importance for:
10^{-3}	Small ions	Atmospheric electricity
10^{-3} up to about 10^{-1}	Large ions, "Aitken" nuclei	Water vapor condensation (Atmospheric electricity)
10^{-1} to 1	Large nuclei	(Water vapor condensation) Optical phenomena in atmosphere Cloud physics, air chemistry
1-10 and more	Giant nuclei (large aerosols)	Cloud physics, air chemistry

Charged Aerosols: Ions

The smallest particles belonging to the aerosol are atmospheric "small ions."[7] They are formed by ionizing radiation originating from terrestrial and cosmic radioactivity which moves an electron from the atomic shell of air molecules. Within a very short time (10^{-8}-10^{-6} sec), this free electron attaches to a neutral atom or molecule. Both of these, the positive as well as the negative molecular ion, are subject to addition of further gas molecules, very rapidly transforming them to larger stable complexes (so-called "cluster" ions) which, on the basis of their properties, consist of approximately ten single molecules. If no other suspended particles are present, the existence of the small ions ends by "recombination" (charge equalization during collision of a positive and a negative ion). If aerosols are present, they attach to a fraction of these and thus form charged aerosols known as "large ions."

---

[7]It is possible that even smaller impurity particles, such as those forming in the atmosphere during the decay of radioactive rare gases—radon, thoron and actinon—have only a short lifetime in the monomolecular state and are fixed to larger suspended particles.

If q pairs of small ions are produced per cm^3 and sec by ionizing processes, the following relation applies to the changes of the positively charged small ion concentration with time:

$$\frac{dn^+}{dt} = q - a\, n^+ n^- \tag{7}$$

for aerosol-free air and

$$\frac{dn^+}{dt} = q - a\, n^+ n^- - \eta_0\, n^+ N_0 - \eta_c\, n^+ N_c^- \tag{8}$$

for air with (charged and uncharged) aerosol particles of identical size. The corresponding change in the negative small ion concentration is obtained by substituting the $+$ index by $-$ signs in Equations (7) and (8). n^+ and n^- refer to the concentrations of positive and negative small ions, respectively, N_c^+, N_c^- are the large ion concentrations, and N_0 is the content of electrically neutral aerosols. a represents the so-called recombination coefficient of small ions and the two η's refer to the coagulation coefficient between small and large ions or uncharged aerosols.[8] For typical atmospheric small ions, a has a value of $1.6 \cdot 10^{-6}$ cm^3/sec; the η-values usually are larger by a multiple and depend specifically on the aerosol particle size.

According to Equation (7), equilibrium is established between ion production and recombination in aerosol-free air if dn/dt becomes zero. In that case, if an equal number of positive and negative small ions are present, and if $n^+ = n^- = n_\infty$, we have

$$n_\infty = \sqrt{q/a} \tag{9}$$

and the mean lifetime τ is calculated from the ionization and the number of ions present in equilibrium according to Equation (1):

---

[8] These represent data of probability that two particles will collide and, to the extent to which charged particles (ions) are involved, will neutralize their charges or, during collision with an ion and an uncharged suspended particle, will combine into one particle. The dimension of these coefficients is $cm^3 sec^{-1}$. The explanation for this is that we are dealing with the expression of an event count (dimensionless) per concentration (count per cm^3) and time (sec): $[(1/cm^{-3})sec^{-1}] = [cm^3 sec^{-1}]$. With regard to the theory of the recombination coefficients, see L. B. Loeb (1947), D. Keefe, P. J. Nolan and T. A. Rich (1959) and others.

$$\tau = n_\infty/q \tag{10}$$

Near the surface of the earth, coagulation of small ions with large ions and uncharged aerosols predominates, so that the term an^+n^- in Equation (8) can be neglected. As a result, the equilibrium small ion concentration near ground level is given by:

$$n_\infty = \frac{q}{\eta_0 N_0 + \eta_c N_c} \tag{11}$$

Equation (11) correlates the equilibrium small-ion concentration with the degree of ionization and the number of aerosol particles. It is exactly valid only for completely monodisperse aerosol, but in the presence of an extremum in the spectrum, it can also be utilized for approximate estimates of polydisperse aerosols. Under these conditions, the mean lifetime of the small ions according to Equation (10) amounts to:

$$\tau = \frac{1}{\eta_0 N_0 + \eta_c N_c} \tag{12}$$

and decreases with increasing aerosol and large-ion concentration.

The size determination of charged particles is based on their behavior under the influence of an electrical field, which induces them to move in the direction of the field or opposite to it depending on their charge. At the same time, they assume a constant end velocity proportional to the driving force, so that they move without acceleration. The explanation of this phenomenon is that motion takes place not in a vacuum but in a so-called *viscous medium.*[9]

---

[9] The actual motion develops by the superposition of two kinetic processes: by participating in thermal molecular motion (Brownian motion), the particle, just like a gas molecule, undergoes zig-zag movements. The respective path-length between two collisions with molecules (or other particles) is not linear in the presence of a superimposed force but becomes a "ballistic parabola." Consequently, a directional effect is imposed on the random thermal motion. Since a new ballistic parabola begins after each collision, no overall accelerated motion can result. Rather, a uniform unaccelerated particle motion results as the integral over the individual path segments. Consequently, the actual motion must depend not only on the size, shape and density of the particle but also on the pressure, temperature and nature of the gas in which motion takes place.

The relation between the driving force and velocity can be described by an equation with certain assumptions concerning the nature of the moving particles. Assuming that *spherical* particles of *density* 1 are involved, the well-known Stokes law applies, which can be written as follows with the modification of E. Cunningham and R. A. Millikan (term in parentheses):

$$K = 6\pi\eta r \cdot v \left\{1 + \frac{L}{r}\left(0.864 + 0.290\cdot\exp\left[-1.25\frac{r}{L}\right]\right)\right\}^{-1} \quad (13)$$

where K = driving force
$\quad\quad\;\; \eta$ = dynamic viscosity of air (at NTP,
$\quad\quad\quad\quad \eta = 1.72 \cdot 10^{-4}$ g cm^{-1} sec^{-1})
$\quad\quad\;\;$ r = particle radius
$\quad\quad\;\;$ v = stationary end velocity of particle
$\quad\quad\;\;$ L = mean free pathlength (at NTP, L = $0.65 \cdot 10^{-5}$ cm).[10]

The Stokes law with its expansion applies to spherical particles with radii of between about 10^{-7} and 10^{-3} cm. The expression in parentheses tends toward a value of unity for $r \gg L$ and to $1.145 \cdot L/r$ for $r \ll L$.

If force K is expressed by e $\cdot$ E (e = particle charge, E = field strength), Equation (13) furnishes the velocity v of the particle in the electrical field. If the latter is divided by the electrical field strength, we obtain the mobility k of the respective particle:

$$k = \frac{c}{6\pi\eta}\left\{\;\ldots\;\right\}\cdot\frac{1}{r} \quad (14)$$

where $\{\;.\;.\;.\;.\}$ represents the parenthetical expression of Equation (13).

Figure 14 shows a diagram for the determination of the radii of charged particles (ions) from their mobility (H. Israel, 1957). The figure is based on Relation (14) and has been calculated with the assumption of spherical particles of density 1 carrying one elementary charge. The reciprocal value of the mobility k, $\tau' = 1/k$, has been selected for the abscissa.

The ionic mobilities are determined by ionometry, *i.e.*, by analysis of current-voltage curves measured by means of flow

---

[10] See the appendix for the calculation of the settling velocity of particles.

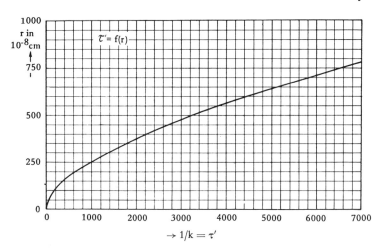

Figure 14. Diagram for the determination of the size of electrically charged particles (ions) from their mobility. Size range of r = 10^{-8} to r = 10^{-5} cm.

capacitors traversed by a stream of ion-containing air (see H. Israel, 1957).

The applicability of this method for the determination of aerosol size spectra is limited in the direction of small as well as large particles. Toward small particles, the probability that they will become electrically charged by the addition of small ions decreases markedly. On the other hand, with increasing particle size, the probability of carrying more than one elementary charge per particle increases.

Both limits can be derived from the equilibrium conditions for the small ion-aerosol mixture: thus, with certain assumptions (Boltzmann equilibrium, equal number and mobility of positive and negative small ions), D. Keefe, P. J. Nolan and T. A. Rich (1959) found the following expression for the ratio Z/N_O of the total number of particles Z per unit of volume to the number N_O of uncharged particles in a homogeneous aerosol (r = particle radius):

$$Z/N_O = K\, f(r) \cdot \sqrt{r} \qquad (15)$$

Constant K has a value of 1044 $cm^{-1/2}$ at room temperature (17°C). The correction $f(r)$ becomes unity for $r > 2 \cdot 10^{-6}$ cm.

Figure 15 shows the trend of Relation (15) for the range of the smallest particles up to 0.16 μm radius with confirming data of P. J. Nolan and E. L. Kennan (1949) obtained with platinum aerosols.

The following relation is obtained for the charge distribution:

$$N_x = 2 N_0 \exp\left(- \frac{(x\,e)^2}{2\,r\,k\,T}\right) \tag{16}$$

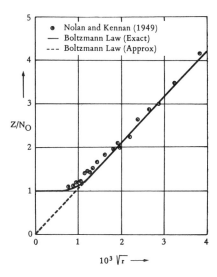

Figure 15. Ratio of Z/N_0 as a function of the aerosol particle radius according to D. Keefe, P. J. Nolan and T. A. Rich (1959).
Abscissa: $10 \sqrt[3]{r}$

where x = 1, 2, 3 . . . represents the number of elementary charges per particle, N_0 is the number of uncharged particles, N_x the number of particles with x elementary charges, e is the elementary charge, r the particle radius, k the Boltzmann constant and T is the absolute temperature.

As an illustration, the charge spectra and values of Z/N_0 are compiled in Table V for a few particle sizes between 0.01 and 1.0 μm diameter and are plotted in Figure 16 for three particle sizes.

In summary, we can therefore say that the ionometric method of aerosol investigation in the range of small and very small aerosols is well-suited for a size determination as long as multiple charges are not to be expected or can be numerically neglected. This is the case for particle sizes up to 0.05-0.06 μm diameter. An extrapolation from a charge count to the total number of aerosol particles present becomes increasingly uncertain toward the smallest dimensions because of the large correction factors.

However, this difficulty can be partly circumvented if the aerosols are charged with an artificial unipolar small-ion current before they are conducted to the flow capacitor (Whitby, Clark, 1966).

Table V

Charge Distribution and Values of the Z/N_o Ratio as a Function of Particle Size*

Diameter in μm	No. of elementary charges per particle											Z/N_o
	0	1	2	3	4	5	6	7	8	9	10	
0.01	0.993	0.007										1.007
0.015	0.955	0.045										1.047
0.02	0.900	0.100										1.111
0.03	0.763	0.236	0.001									1.316
0.06	0.550	0.430	0.020									1.818
0.1	0.424	0.48	0.09	0.006								2.359
0.3	0.241	0.41	0.232	0.093	0.024	0.005						4.150
1.0	0.133	0.253	0.214	0.162	0.109	0.065	0.035	0.017	0.007	0.003	0.001	7.518

*According to D. Keefe, P. J. Nolan and T. A. Rich (1959).

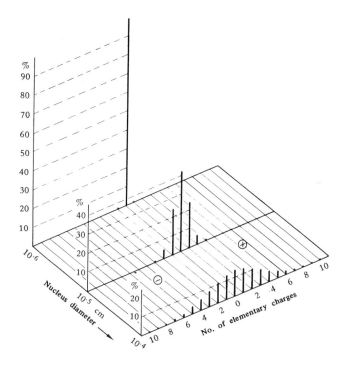

Figure 16. Distribution of aerosol particles in per cent with 0, 1, 2 . . . elementary charges for particle sizes of 0.01, 0.1 and 1.0 μm according to D. Keefe, P. J. Nolan and T. A. Rich (1959).

Whitby succeeded in measuring aerosol spectra and concentrations in the size range of 0.015-1.0 μm by ionometry.

Water Vapor Condensation on Aerosols

One of the most important properties of aerosols which at the same time represents a substantial broadening of possibilities to observe particles in the submicroscopic size range is their capacity under suitable conditions to attract water vapor from their environment for condensation on them. According to this finding which dates back to M. Coulier (1875), the condensation of water vapor in the atmosphere requires the presence of particles that can serve as condensation nuclei. Detailed studies of these "condensation nuclei," their properties and their activity began with the voluminous investigations of J. Aitken (1887/88) and still represent the most important partial problem of the physics of small and very small aerosols (see, for example, H. Landsberg, 1938, and others).

The condensation process can be described in the following way. If a water droplet is in an atmosphere saturated with water vapor, for example, in a closed chamber over a plane water surface, a higher vapor pressure will be present above it due to surface curvature than above the plane water surface. The relation is given by the expression[11] derived by W. Thomson (1871):

$$\ln \frac{p'}{p} = \frac{2\sigma}{r\rho_0 RT} \qquad (17)$$

where p' = vapor pressure over the curved water surface
p = vapor pressure over the plane water surface
σ = surface tension
r = droplet radius
ρ_0 = density of water
R = universal gas constant
T = absolute temperature

This relation states that water droplets cannot continue to exist next to a plane water surface since they must evaporate and release their water vapor to the plane surface. In other words,

---

[11] For mean temperatures, the approximation

$$\Delta p = \frac{1}{r} \cdot 0.191 \cdot 10^{-4} \text{ mm Hg}$$

results for the relation between $\Delta p = (p' - p)$ and the particle radius r (expressed in mm).

droplets can exist only in a water vapor-supersaturated atmosphere that is in equilibrium with their evaporation tendency.

The values of this equilibrium supersaturation increase with decreasing droplet radius in the manner described by Curve I in Figure 17.[12] These conditions become substantially different when the droplets carry an electrical charge or contain a contaminant. In both cases the vapor pressure over their surface decreases, so that their existence becomes possible even in a saturated and unsaturated atmosphere.

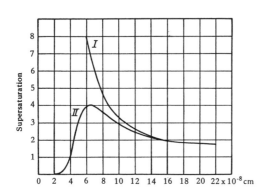

Figure 17.   Equilibrium supersaturation for droplets of radius r. Curve I: uncharged droplets; Curve II: droplets with one elementary charge.

The corresponding relationships result from a modification of the Thomson Equation (17), which is replaced by:

$$\ln \frac{p'}{p} = \frac{1}{\rho_0 \, R \, T} \cdot \left(\frac{2\sigma}{r} - \frac{e^2}{8\,\pi\,r^4} \right) \tag{18a}$$

for droplets of charge e and

$$\ln \frac{p'}{p} = \frac{2\sigma}{\rho_0 \, R \, T \, r} - \frac{H' \rho_k \, r_k^3}{\rho_0 \, r^3} \tag{18b}$$

for droplets in solution. In Equation (18a), e is the droplet charge and in Equation (18b) H' is an empirical value known as the hygroscopic factor (see International Critical Tables *3*, p. 292); ρ_k and r_k are the density and the radius of the soluble nucleus present in the droplet, for example, of a salt crystal or the impurity contained in it.

---

[12] "Auto nucleation" in a highly supersaturated atmosphere is not discussed here.

The influence of an electrical charge of the droplet, defined by Equation (18a), is illustrated by Curve II in Figure 17. It becomes noticeably effective only for the smallest particles and offers the basis for visualization of small ions in a Wilson cloud chamber operating at suitable supersaturation. With increasing droplet size, the charge influence decreases rapidly and as a rule can be neglected compared to the influence of soluble impurities.

Equation (18b) furnishes a relation between the equilibrium size of a droplet with a certain content of soluble impurity substance at a given relative humidity. At the same time, the equation indicates the supersaturation that must be reached so that the respective solution droplet can grow to visible size by water uptake.

According to H. L. Wright (1936), Equation (18b) can be substituted by the approximation:

$$\frac{F}{100} = e^{P/r} - \frac{Q}{r^3} \tag{19}$$

where $F = 100 \cdot p'/p$ is the relative humidity, P is the expression on the right side of Equation (17), multiplied by r, and $Q = H' \cdot M$ (M = solution concentration in mol per liter). The hygroscopic factor H' is about $4 \cdot 10^{-2}$ for NaCl, HNO_2 and H_2SO_4 solutions and about $8 \cdot 10^{-2}$ for $MgCl_2$ solution.

Figure 18 shows the growth curves and the supersaturation required for growth for the size range of $r = 10^{-7}$ to $r = 10^{-4}$ cm as calculated by H. L. Wright (1936) according to Equation (19). Accordingly, a droplet with a given impurity content will adjust to an equilibrium value by water uptake as a function of the relative humidity in its vicinity, *i.e.*, it grows with increasing humidity until it has reached and exceeded the culmination point in the supersaturated region after the water vapor saturation is exceeded. From then on, it must continue to grow by further water uptake. This process of condensation is known to remain basically the same when the condensation nucleus consists of insoluble material, provided it is small enough (C. E. Junge, 1936).

Since the method was first described by J. Aitken (1890/91), condensation nuclei are counted with utilization of their growth by condensation in such a manner that the nucleus-containing air is humidified up to saturation and is subsequently supersaturated up to 100% and more by adiabatic cooling. The nuclei growing into droplets in this process drop by gravity on a counting slide

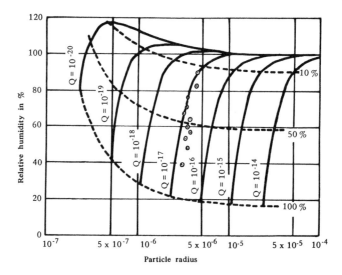

Figure 18. Particle size of solution droplets in relation to the relative humidity and corresponding supersaturation for condensation at different Q-values according to H. L. Wright (1936). The individual curves start at saturated solution (100%); dashed curves indicate the corresponding solution concentrations. The middle curve has been constructed through test points.

and are then counted in a dark field. In more recent versions, the light attenuation by the fog forming during expansion is observed in place of making a direct count (L. W. Pollak and T. Murphy, 1952, and others).

The supersaturation necessary for such nucleus counters is still not fully known. As a rule, it is assumed to be unusually high. Thus, according to J. Aitken, nuclei formed by annealing require up to 200% for condensation to take place.

In contrast, more recent studies of the condensation process show that it actually takes place in a manner substantially different than that described above and that the theoretical supersaturations are *not* necessary. W. Wieland (1955) and G. Gotsch (1962) found that when a different measuring principle is used in which supersaturation is maintained for some time ("mixed cloud counter"), the same number of nuclei are found with only a few tenths of a per cent of supersaturation, as with an expansion counter in which the supersaturation was calculated to be 100% and more from the expansion ratio.

In confirmation of this, H. Israel and N. Nix (1969) found recently that the process also has a significantly different course in the expansion counter than was assumed up until now: condensation begins *immediately* with the start of expansion rather than only after a high supersaturation is reached. The calculated supersaturations thus do not even develop.

Furthermore, H. Israel and N. Nix (1969) and N. Nix (1969a, 1969b) developed a new measuring method that allows observation of the condensation process on a single particle under precisely known thermodynamic conditions. They arrived at the result that growth of the nuclei into visible droplets takes place more rapidly by at least two orders of magnitude than is calculated, for example, by diffusion theory with the use of Equation (18b). Furthermore, they found that droplet growth takes place only as long as the supersaturation S increases in the surrounding medium ($dS/dt > 0$).

Figure 19 shows an example of the condensation and reevaporation process on platinum nuclei of $5 \cdot 10^{-7}$ cm diameter. The system was selected in such a way that the particle moves on a small ellipse with semiaxes of about 10 and 50 μm at the rhythm of a periodic pressure variation that produces a humidity variation of between 96 and 104% in the medium surrounding the particle. In Figure 19a, the shaded area on the left shows the supersaturated and the area on the right the undersaturated region. The nucleus becomes visible soon after the start of supersaturation, then grows to the maximum supersaturation. After this is exceeded, it begins to evaporate again still in the supersaturated range, finally becoming invisible again at the right top in the undersaturated zone. Figure 19b shows the microphotograph of such a process. The nucleus becomes visible with a size of $2 \cdot 10^{-5}$ cm at about 101.3% humidity and grows to about $4 \cdot 10^{-4}$ cm until maximum supersaturation of 4% is reached. The brightness of the track is a criterion of the scattered light intensity and thus of the particle size. Figure 19c shows the same process with a time scale recorded with a stroboscope (distance of light pips = 2 msec).

The total process remains qualitatively the same when the time in which the entire cycle is completed (120 msec in the figure) is reduced to 20 msec (limit of detectability at this time). The only difference is that the final size reached by the droplet decreases by about half.

The discrepancy between the above sketched "classical" concept concerning the condensation process and its real course

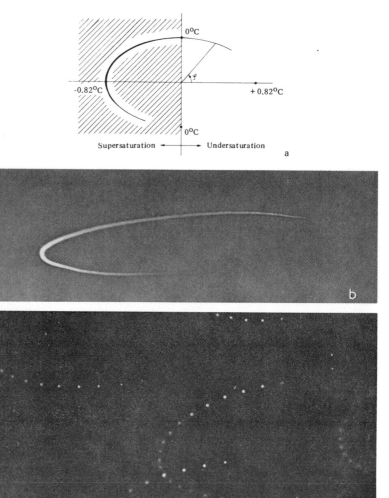

Figure 19. Photographs of the condensation and evaporation cycle in its temporal change.

a. Diagram of recording technique. Variation of humidity between 104% (left reversal point) and 96% (right reversal point).

b. Photograph of a platinum particle (initial size about $5 \cdot 10^{-7}$ cm; start of visibility at about $2 \cdot 10^{-5}$ cm and 101.3% relative humidity); duration of total cycle 120 m/sec. Maximum particle size about $4 \cdot 10^{-4}$ cm.

c. The same with a stroboscopically recorded time scale (3 msec period between two light flashes).

requires a revision of old ideas. The starting point for this may be found in the argument that the transport of water vapor to the droplet surface cannot be explained simply by diffusion. Rather, it must be assumed that the diffusion-related transport of H_2O molecules to the particle, which probably falls approximately into the order of magnitude of the mean free pathlength of H_2O vapor, is slower than the statistical gas-kinetic impingement of H_2O molecules on the droplet surface, which naturally leads to a much more rapid droplet growth.

If supersaturation increases ($dS/dt > 0$), additional water vapor is released for condensation within a zone having the thickness of the mean free pathlength and this no longer needs to be transported to the droplet from outside by the slow diffusion process. This quantity adds practically instantaneously to the drop surface. Since no supersaturation can develop in the immediate vicinity of the droplet as a result of this process, growth is possible only as long as supersaturation *increases* outside of this region (see footnote 8, p. 33).

Thus, the condensation process takes place in the form of diffusion only in the more remote zone surrounding the condensing surface, while it is a process of gas kinetics in its immediate vicinity. The former process is determined by supersaturation S as such, while the latter is controlled by its increase with time.

Size Spectrum of Aerosols[13]

In spite of considerable differences in the mean aerosol concentration from one locality to another and its fluctuation in connection with meteorological influences (see "Meteorology and Climatology of Aerosols," page 69), their size spectrum is remarkably similar and consistent. For continental aerosols, the studies of C. Junge (1953, 1963) show that this distribution can be represented as shown in Figure 20 in slightly idealized form.

The spectra show a maximum population in the size range between 0.01 and 0.1 μm, which decreases toward smaller and larger particle sizes. The decrease in the direction of larger particles can be represented by a power distribution law formulated by C. Junge having the following form: If $N(r)$ is the number of particles per unit volume belonging to the radius range of $d(\log r)$ and dr, respectively, the frequency distribution can be described by the relations:

---

[13] See also "Chemistry of Aerosols," page 55.

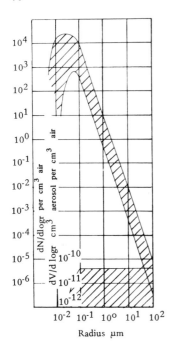

Figure 20. Size distribution of continental aerosols according to a summary representation of C. Junge (1967). The shaded zone shows the range of variation. In the lower part of the figure the corresponding distribution of aerosol volumes is shown.

$$\frac{dN(r)}{d\,(\log r)} = c \cdot r^{-\beta} \tag{20a}$$

and

$$\frac{dN(r)}{dr} = 0.434 \cdot c \cdot r^{-(\beta + 1)} \tag{20b}$$

As a rule, a value of about 3 is obtained for the exponent β; c is a constant with a value falling between 1 (high-mountains) and 10 (lowlands), referred to 1 cm^3, if r is expressed in μm.

If the total volumes of the aerosols are determined as a function of size according to the "Junge Law" represented by Equation (20), we obtain the trend shown in the shaded lower part of Figure 20. Thus, starting with a size of about 0.1 μm and more, the total volume per unit size range remains unchanged regardless of particle size.

In the lower part of the aerosol spectrum (size range between about $5 \cdot 10^{-3}$ and 10^{-1} μm particle radius), our knowledge concerning the size distribution is still relatively scant. For example, the question of whether the "population" of the size spectrum is also monotonous here or is resolved into discrete single groups cannot yet be definitely answered. Size determinations by ionic spectroscopy (H. Israel and L. Schulz, 1932) admittedly suggest the presence of "line spectra," *i.e.*, discrete groups of particle sizes, but attempts to determine size distributions in this range with other measuring methods (for example, A. L. Metnieks and L. W. Pollak, 1962 as well as S. Twomey and G. T. Severynse, 1964) rather justify the assumption that a continuous spectrum also exists in this range. Because of the remaining uncertainties, the range of small aerosols has not been included in the Junge representation of Figure 20.

In marine aerosols, the conditions differ insofar as the small aerosols are much less abundant. Furthermore, the range of maximum frequency has been shifted toward larger radii by about one order of magnitude. From then on, these aerosols also follow the Junge law in their frequency distribution (for further details, see page 57).

The consistencies in the aerosol size spectra discovered by C. Junge suggests that their development can be explained by certain processes of gas kinetics taking place in the air-aerosol mixture. After their penetration into the atmospheric gas space, the aerosol particles are subject to Brownian motion. This results in collisions between individual particles, giving rise to coagulation. These diffusion phenomena also lead to the attachment of suspended particles on walls or the addition of small particles to larger ones. Finally, with increasing particle size, their settling velocity gradually becomes more important and produces separation by sedimentation. As an example Figure 21 shows the behavior of an aerosol in a closed space.

A decrease of the number of particles with time occurs, taking place at first approximately in proportion to the square of their number if the latter is large, slowing down subsequently, and gradually changing to a reduction that is proportional to the number itself, *i.e.*, exhibiting a trend following Equation (21):

$$\frac{dn}{dt} = -Kn^2 - \lambda n \tag{21}$$

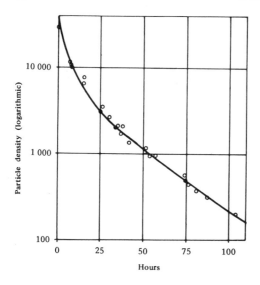

Figure 21. Decrease in number of nuclei with time in a gasometer of 330 l volume charged with natural room-air aerosol (according to a measurement of P. J. Nolan, 1941). Circles: test points; solid line: theoretical trend according to the integral of Equation (21).

Integration of (21) results in the solid line of Figure 21.[14]

This experiment shows that when the particle number decreases, two different processes are evidently superimposed. The term $-Kn^2$ on the right side of Equation (21) describes the particle loss by collisions and subsequent coagulation according to the well-known coagulation theory of M. V. Smoluchowski (1916). The second term $-\lambda n$ represents the particle loss due to diffusion toward vessel walls and possibly other large surfaces relative to the particles, for example, those of coarse suspensions. This component may also contain losses by sedimentation.

The following basic principles of gas kinetics can serve as a starting point for an understanding of the specific processes and their mathematical treatment. For a given particle size, coagulation, diffusion and sedimentation are related in the following simple manner. If we define the "mobility" b of the particles according to Equation 13 by their velocity v, which they receive under the influence of a force 1, the following relations are valid:

For the falling speed:

$$v = b\,m\,g \tag{22}$$

---

[14] In the present case, the following values were determined for the constant K and λ:

$$K = 1.9 \cdot 10^{-9} \ cm^3 sec^{-1} \quad \text{and}$$
$$\lambda = 7.9 \cdot 10^{-6} \ sec^{-1}.$$

b = mobility in sec · g^{-1}
m = particle mass in g
g = acceleration due to gravity in cm · sec^{-2}

For the diffusion coefficient D:

$$D = \frac{R\,T}{N} \cdot b \qquad (23)$$

D = diffusion coefficient
R = gas constant
T = absolute temperature
N = Avogadro number

and for coagulation:

$$\frac{dn}{dt} = -Kn^2; \quad \text{with} \quad K = -8\pi D\,r \quad \text{for} \quad L \ll r \qquad (24)$$

r = particle radius
K = coagulation constant.

By determining one of the three parameters b, D or K, it is thus possible to control the processes of coagulation, diffusion and sedimentation in a homogeneous aerosol.[15],[16] If the particle

---

[15] An important application of this appears in the analysis of small aerosols. In order to determine their diffusion coefficient as a characteristic quantity, the aerosol-containing air is passed through a so-called diffusion battery consisting of one or several narrow channels. The numerical particle loss that occurs during flow-through results in the value of the diffusion coefficient D. For details, see A. L. Metnieks and L. W. Pollak (1962) and others. For the ratio n_e/n_a of the particle concentration n_e in the air after its discharge from the diffusion battery to the original concentration n_a, theory furnishes the following relation for a homogeneous aerosol:

$$n_e/n_a = 0.9099 \cdot e^{-x} + 0.0531 \cdot e^{-11.369x} \qquad (25)$$

with

$$x = 3.77 \cdot \frac{h\,l\,p}{b\,Q} \cdot D \qquad (25a)$$

h = altitude ⎞
l = length   ⎬ of the vertical rectangular diffusion channel
b = half-width ⎠
p = number of parallel flow channels of identical type
Q = aspirated quantity per sec
D = diffusion coefficient

mixture consists of particles of different size, as is usually the case, the coagulation Equation (14) must be generalized as follows: between two particle groups 1 and 2 with diffusion coefficients D_1 and D_2, radii r_1 and r_2, and concentrations $n(r_1)$ and $n(r_2)$ (number of particles per unit volume), we obtain Relation (26) for the coagulation rate $dn(r_1, r_2)/dt$:

$$\frac{dn(r_1, r_2)}{dt} = -K(r_1, r_2)\, n(r_1)\, n(r_2)\, dr_1\, dr_2 \qquad (26)$$

where $K(r_1, r_2) = 4\pi(D_1 + D_2)(r_1 + r_2)$ for $L \ll r$.

G. Zebel (1966), with consideration of a "gas-kinetic correction factor" for $L \gtrsim r$, calculated numerical values of the "coagulation constant" $K(r_1, r_2)$, which are compiled in Table VI for a few combinations of particle radii.

Table VI

Coagulation Constant K (10^{-10} cm^3/sec) for a Few Combinations of Particle Radii*

r_2	0.001	0.01	0.1	1.0
r_1				
0.001	8.78			
0.01	180.2	21.0		
0.1	8845	168.5	11.10	
1.0	178100	2023	35.95	6.44

*According to G. Zebel, 1966. Radii in μm.

This compilation shows first of all that the probability of coagulation between particles of different sizes is considerably greater throughout than the probability of coagulation between particles of equal size. Secondly, the coagulation effect decreases rapidly

(Footnote 15 continued)
Laminar flow, absence of electrical forces, adhesion of particles contacting the wall and minimum pressure differences in the diffusion battery are prerequisites for the validity of Relation (25).

[16] For the numerical values of the settling velocity of aerosol particles, see the appendix.

from the very small aerosols to the larger ones. Both factors together lead to spectral changes in a particle mixture of such a type that a shift toward larger dimensions occurs.

This process of "aging" of a given aerosol has been investigated theoretically by several authors. G. Zebel (1958) determined the change of an aerosol in the range of small particles with an initially relatively narrow size spectrum, when the aerosol is allowed to remain in a closed chamber, with the result as shown in Figure 22. C. Junge (1955) made similar calculations for a natural continental aerosol with a correspondingly broad size spectrum and obtained the results shown in Figure 23.

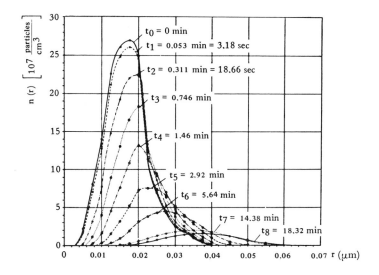

Figure 22. Variation with time of the size spectrum of a small-particle aerosol of narrow size distribution in a closed space according to G. Zebel (1958).

Both figures show the coagulation-determined tendency in the aerosol size spectrum to shift toward larger particles. In the first case, a reduction in total particle number occurs as expected, while in the natural aerosol an adaptation to the power distribution law of Figure 20 takes place under the influence of the transformation process. The smaller the particles of the initial distribution, the shorter will be their expected lifetime.

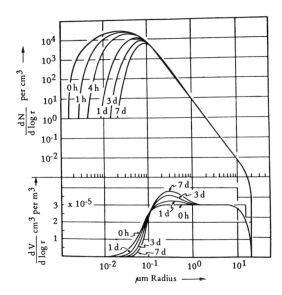

Figure 23.   Variation with time of the size distribution and total volume of a continental aerosol according to C. Junge (1955). h = hours, d = days.

Thus, a distribution of the type sketched in Figure 20 gradually develops in an aging aerosol where the lower limit falls at about 0.005-0.01 μm, while a zone of maximum population density develops approximately in the range between 0.01 and 0.1 μm.[17]

This coagulation-determined mass transport from the small nucleii to larger agglomerates has only an insignificant influence on the power distribution in the range of larger particles, since the contribution of the "small" total aerosol volume remains minor (see lower part of Figure 22). On the other hand, the constancy of the aerosol volume in the range of the Junge law suggests that the numerical dependence on particle size again is determined by coagulation (proportionality with r^{-3}).

The question of how the distribution develops according to the Junge law has not yet been elucidated to satisfaction. S. K. Friedlander (1960), as well as R. E. Pascerie and S. K. Friedlander (1965), have assumed that the distribution is formed by the

---

[17]It is interesting that this maximum interval coincides with the size range in which the always highly populated spectral interval of the large ions is found in the ionic spectrum (see Table IV).

coagulation-related transport of nuclear mass from small to larger particles as described, leading to a "self-preserving distribution."

C. Junge (1955, 1963, 1967) believes that other processes forming the spectrum are more probable for various reasons (primarily because of the too slow transformation process in the aerosols compared to the duration of meteorological influences), and he considers processes of direct aerosol production in the region in question as well as influences of the condensation and evaporation process of clouds and raindrops as controlling factors.

Optical Effects of Aerosols ("Haze Optics")

The optical characteristics of the atmosphere are subject to substantial changes due to the presence of suspended particles. These are based on scattering and absorption of light during transmission through aerosol-containing air. As a whole, they act by attenuating and modifying the sunlight penetrating the atmosphere and varying the visibility range. Furthermore, they have more specific effects resulting from the dependence of absorption on the wavelength of light and on the aerosol size distribution, which become manifest in the form of changes in the blue color of the sky as well as some occasional color phenomena—discoloration of sun and moon, "halo."

A study of these relationships first of all serves to interpret the mentioned phenomena. At the same time, it offers further possibilities for the determination of certain aerosol properties by optical methods. The well-known Mies theory of light-scattering on small particles serves as the basis for the problems of haze optics. A detailed description of this theory in its special application to atmospheric aerosols can be found in the paper of F. Volz (1956) in *Handbuch der Geophysik*. A few particularly important results for aerosol studies will be briefly discussed.

The blue color of the sky results from scattering of sunlight on air molecules, the intensity of which is inversely proportional to the fourth power of the light wavelength λ, according to Rayleigh. In contrast, the light scattered by fog and cloud droplets is white, *i.e.*, it shows no spectral change. In the size range between molecular dimensions and the cloud elements of about 10-20 μm, therefore, a change occurs from the λ^{-4} to the λ^0 law which allows us to expect a relationship with the size of the scattering particles.

On the basis of measurements of solar radiation, A. Ångström (1920/30) derived a wavelength dependence of haze scattering according to λ^{-a} and found an a-value of 1.3 that has proved to be useful since that time to describe the attenuation of radiation for visibility ranges of 1-100 km as well as for different climates. The fact that such a fairly generally valid wavelength exponent is obtained in spite of the mentioned size spectra of atmospheric aerosols can be explained by the applications of the Mies theory to the generally predominating size distribution reflected by the Junge law as illustrated by Figure 20. On this basis, and with the assumption of spherical particles with a refractive index of 1.5, the calculated scattering shows a wavelength dependence in proportion to $\lambda^{2-\beta}$, where β is the exponent of the Junge law of Equation 20. In performing the calculation, we find that the different aerosol size ranges contribute to the total absorption in very different ways: for particles smaller than 0.1 μm, the scattered fraction that is proportional to the square of the radius decreases rapidly. Particles larger than 1 μm diameter practically no longer contribute to scattering because of their rapidly decreasing number. Thus, generally only particle sizes between about 0.1 and 1 μm become effective in the wavelength exponent (2-β), *i.e.*, particles in the vicinity of the maximum of the size spectrum.

The agreement between the Junge value of $\beta = 3$ (for which limits of 2.5-3.5 are probable) and the value of $\beta = 3.3$, which can be assumed on the basis of the Ångström measurements, is remarkably good. It is a reciprocal confirmation of two findings that were obtained by entirely different methods. In particular, we can conclude from this that the Junge power law of the aerosol size distribution can evidently lay claim to having world-wide character. A relationship with the same tendency is found between atmospheric visibility and light scattering by aerosols.

The visibility is defined as the distance at which the brightness contrast between a target (brightness B_z) and the horizon (brightness B_h) disappears. This can be described by the relation:

$$B_z = B_h (1 - e^{-\sigma z})   \qquad (27)$$

σ = (total) scattering
z = distance

With the use of the empirical value according to which the limit of visibility has a value of $(B_h - B_z)/B_h = 0.02$ (contrast threshold), this relation is transformed into the following expression for the visibility V:

$$V = \ln \frac{1}{0.02} \cdot \frac{1}{\sigma} = \frac{3.912}{\sigma} \qquad (28)$$

The quantity of σ contains the "Rayleigh scattering," the scattering produced by aerosols and possibly gas absorption. Since these contributions can be easily separated, the visibility and its change also offers information about σ (for details see C. E. Junge, 1963, F. Kasten, 1967, and others).

Calculations of the scattering function as a function of the scattering angle (angle between direction of illumination and line of sight) for aerosol spectra corresponding to the Junge power distribution of Equation (20) lead to the interesting result that the scattered light distribution is the same for all wavelengths. In other words, no unusual scattering-related color phenomena can occur in this case (see E. De Bary and K. Bullrich, 1962). Measurements of the scattering function in an aerosol near ground level largely confirm the power law with a β-value close to 3 (K. Bullrich, 1960). On the basis of the scattered light distribution in the zone surrounding the sun, F. Volz (1956) concludes that the power distribution law is not followed.

The decrease of the wavelength exponent from -4 for Rayleigh scattering to about -1.3 for haze scattering is an explanation of the observation that haze changes the blue of the sky in the direction of white, and that the sun and moon can change their color up to dark red during the rising and setting phases. Furthermore, the wavelength dependence of σ has the result that visibility in a hazy atmosphere is greater in the red than in the blue region.

Anomalous color phenomena caused by scattering can occur only with deviations from the power distribution law. These include the "halo," a brownish ring around the sun that has been explained by volcanic dust in high layers of the atmosphere, the phenomena of the "blue sun" of 1950 which were caused by aerosols from Canadian forest fires (see R. Wilson, 1951 and H. Runge, 1951), and finally, the various forms of colored smoke (cigarette smoke, sulfur fumes). Details concerning the aerosol distributions that are the basis of these color phenomena can be found in the studies of F. Volz (1956) and others.

CHEMISTRY OF AEROSOLS

The question of the chemical nature of atmospheric particulates has become accessible to systematic research only in recent years.

The reason for this resides primarily in the difficulties of chemical analysis of extremely small quantities of substance, which even today have been only partly overcome.

The first possible tool consists of chemical analysis of precipitation and its residues, since this process produces an automatic enrichment of the suspended material. However, retrospective conclusions from such analyses concerning the original state of the condensation-favoring substance cannot always be considered reliable in view of the possibility of reactive modification of these components in aqueous solution or suspension. Another method is to consider questions of chemical composition of the aerosols to be largely equivalent to questions concerning their sources, production and development under atmospheric influences. By combining both possibilities, a fairly clear picture of aerosol chemistry has been obtained, at least in its basic outlines and this will be described briefly here.

The *sources of aerosols* are located almost exclusively in the terrestrial region. They are predominantly related with processes on the surface of the earth. Aerosol particles of extraterrestrial origin exist also, but quantitatively they play no role in the atmospheric aerosol balance.

The processes leading to the *production of aerosols* can be summarized in the form of three groups:

1. Combustion processes of all types are a main source. They introduce combustion products of the most diverse type and size into the atmosphere. Depending on environmental conditions, natural sources (forest, brush and grass fires, volcanic activity) or anthropogenic sources (heating, industry, motor vehicle traffic) may predominate.

2. Formation of suspended particles by reactions between trace gases in the atmosphere, taking place under the influence of heat and radiation in the presence of atmospheric water vapor, is another source. Examples, among others, are the formation of ammonium chloride (NH_4Cl) from NH_3 and HCl or the oxidation of SO_2 into SO_3, with the resulting formation of sulfuric acid (H_2SO_4). Again, air pollution due to civilization carries a notable weight here.

3. Introduction of finely dispersed material of continental and oceanic origin into the atmosphere by wind and exchange processes. This group also includes materials of organic origin such as pollen and microorganisms (spores, bacteria) as well as small insects and insect fragments.

It may be expected that depending on the nature and course of these processes, their aerosol yield will have a preferential effect on different ranges of the total size spectrum.

The *development of aerosols under atmospheric influences* concerns the modifications of their size distribution by evaporation of primary particles, condensation of water vapor with changes in humidity in their vicinity, and by coagulation. Furthermore, chemical reactions in the air between particle substance and atmospheric gases or trace gases become of significance here.

It seems indicated to classify aerosols by their origin and to discuss first the two main regions of origin—the ocean and continents—with regard to their aerosol production in order to characterize the aerosols derived from them in their chemical properties. As a supplement to this, aerosols of extraterrestrial origin will also be considered briefly at the conclusion of this section.

Aerosols of Marine Origin

The oceans represent the largest natural aerosol source of worldwide extent. The parent material of the aerosol produced by them consists of sea salts, primarily NaCl.

The aerosols are formed from spray penetrating the atmosphere: fine air bubbles expelled from the surface of the water separate the surface and thus produce small droplets that decrease further in size in air by evaporation. Figure 24 shows the process of droplet formation in a schematic diagram. The rising air bubbles burst on the surface of the water. This results in a fine ascending jet of water that soon disintegrates into a number of approximately equal droplets (W), which may be transported to an altitude of up to about 15 cm depending on their size. Their size amounts to about 1/10 of the bubble dimension (C. F. Kienzler *et al.*, 1954). At the same time, a large number of much smaller water particles (M) can be observed in every such

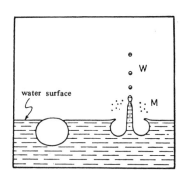

Figure 24. Formation of spray droplets during expulsion of air bubbles from the surface of the water (from C. E. Junge, 1963). W = droplets, M = water particles.

process; evidently these form during disintegration of the thin
water film on the bubble surface (B. J. Mason, 1954).

Both groups of droplets are exposed to air motion and exchange, and are dispersed horizontally and vertically. In this
process their size decreases by evaporation until they reach an
equilibrium determined by the relative humidity, salt concentration and the radius of curvature. At low relative humidity, there
may even be a formation of concentrated salt solutions and beyond this, crystallization of the salt.

A further possibility for the entry of sea salt into the air consists of the removal of particles from wave crests by wind. The
particles formed usually are considerably larger than those discussed above and have only a short lifetime, since they are rapidly
deposited again. They are probably responsible for the occasionally very high salt concentrations of 50-1000 $\mu g/m^3$ of the air in the
immediate vicinity of the coast extending only a few kilometers
into the continent (H. R. Neumann, 1940).

As mentioned briefly in the section on aerosol sizes (page 45),
marine aerosols in the range of larger particles exhibit the same
power distribution as continental aerosols. Figure 25 shows a few
measurements of sea salt particles. A clear difference becomes

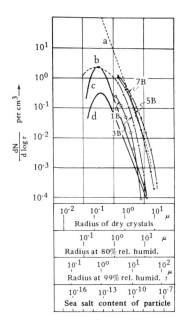

Figure 25. Size spectra of sea salt aerosols. Curves 1B, 3B, 5B and 7B show the results of A. H. Woodcock (1953) at wind speeds of 1, 3, 5 and 7 Beaufort. Curve *a* shows the trend of a continental size spectrum for compari-son; *b* is an extrapolation of the 3B curve; *c* and *d* show values for Dublin and the west coast of Ireland (A. L. Metnieks, 1958). The scales on the abscissa represent the corresponding values for (dry) salt volume, droplet size at 80% and 99% relative humidity and salt content (from C. E. Junge, 1963).

evident in the marked shift to the right of the maximum of the size spectrum, *i.e.,* in the direction of larger particles.

In view of the sparse experimental data available, little can be said concerning the trend of the spectrum in the small particle range and the origin of these particles in the marine aerosol. On the whole, however, the range of Aitken nuclei appears to be more densely occupied over the ocean than would be expected on the basis of the "shift to the right" of the maximum range of sea salt particles according to Figure 25.

The production of sea salt aerosols sketched in Figure 24 indicates that a relation may exist between aerosol production and wind strength over the ocean. This is confirmed by the relationship between the total salt content of the air over the oceans and the wind speed shown in Figure 26. In contrast to this, according to the studies of D. J. Moore (1952), no detectable relation exists between the number of Aitken nuclei and wind speed or wave height.

This discrepancy suggests that the relatively large number of Aitken particles present over the ocean essentially are not of marine but of continental origin and are mixed with marine aerosols by large-scale air and exchange motion. C. E. Junge accounts for this by the assumption that the oceanic aerosol spectrum consists of two components (see Curve 2 in Figure 27).

The idea of broad-scale mixing of aerosols of various origins naturally is equally valid for marine components in continental aerosols. Thus, already decades ago, surprisingly high Cl^- ion

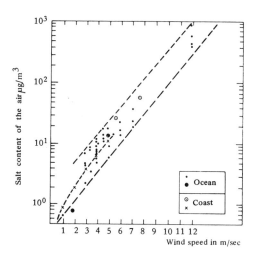

Figure 26. Salt content of the air in $\mu g/m^3$ as a function of wind speed according to various measuring series over the ocean and on the coast (from C. E. Junge, 1963).

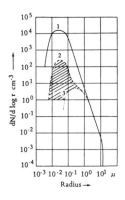

Figure 27. Model spectra of atmospheric aerosols according to C. E. Junge.

Curve 1: continental aerosol;
Curve 2: marine aerosol;
Curve 3: sea salt fraction in marine aerosol.

The shaded area represents the marine aerosol fraction that cannot be explained by the spray effect.

concentrations were observed in the precipitation and thawing hoar frost at a great distance from the coast, for example, on the Sonnblick Mountain. Extensive series of measurements from the fifties produced the picture shown in Figure 28.

According to the above, the Cl⁻ ion concentration at first decreases rapidly in the inland direction from the coast, but then changes to a nearly constant level over the entire continent. This finding seems to indicate that the fraction of marine origin is limited to the vicinity of the coast; this may also be considered probable in view of the relatively large sea salt aerosols. Other sources must be assumed to be present in the interior of the continent (see the following section).

In conclusion, we should note the possibility of biological aerosol production in the marine zone, which was discussed by A. Goetz. To the extent to which anything can be said about this at this time, these involve some water-insoluble end products of biological reactions collecting on the water surface in the form of a fine film. According to A. Goetz (1960) they can be recognized by the well-known "mirror spots" on the water surface, which can be observed in a mild breeze (see Figure 29) and which can be explained by a modification of surface tension produced by these "biofilms." From the surface of the ocean, these materials are then entrained into the atmosphere during the bursting process of fine air bubbles as described above. The aerosol spectrometer (an aerosol centrifuge efficient down to sizes of a few tenths of a μm) has indicated a large number of particles in the lee of such mirror spots. Such biocolloidal admixtures have an inhibiting effect on the condensation process on sea salt aerosols since they hinder water vapor transport to and from the particle surface.

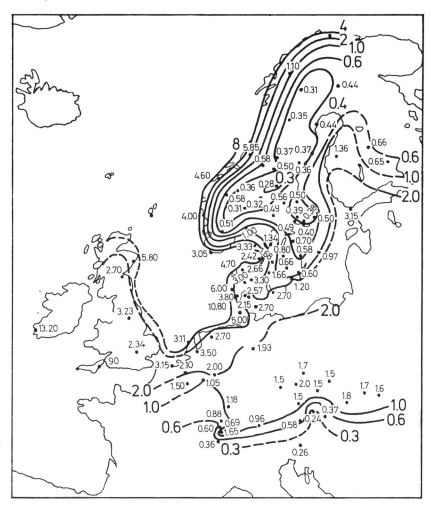

Figure 28. Chlorine ion (Cl⁻) concentration in rain water (mg/1) over Europe (from C. E. Junge, 1963).

Aerosols of Continental Origin

The aerosols of continental origin are incomparably more complicated in formation and composition. From the standpoint of chemical analysis, we distinguish between small aerosols, about which only indirect information can be obtained, and larger particles, which are accessible to direct microchemical analysis.

Figure 29.   "Mirror spots" on the water surface according to A. Goetz (1960).

\Small Aerosols.\ According to Table IV, small aerosols include ions and condensation nuclei of the Aitken range up to a diameter of about 0.1 μm. Their lower size limit probably is above the molecular range at about 10^{-7} cm, as can be deduced from ionic spectra.

The main source of small aerosols, as noted earlier, is all types of combustion processes. In addition, aerosol-forming reactions take place between trace gases under the influence of heat and radiation.

Thus far, no direct information is available on the chemical composition of these particles. However, the following indirect conclusions would seem to be permitted. Since our considerations in connection with Figure 22 indicate that coagulation causes a growth of the material contained in the small particles into larger particles, their composition should allow retrospective deductions concerning the nature of the original material. Furthermore, experience shows that anthropogenic aerosol production leads to the formation of not only the Aitken nuclei but also larger particles that can be analyzed by microtechniques. This, too, permits a conclusion concerning the similarity of the material.[18]

---

[18]As an addition we should note a new possibility for the study of condensation nucleus material. The above-described possibility of the condensation

Figure 30. Examples of condensation cycles in a mixed aerosol of Pt-nuclei and urban aerosol according to N. Nix (1969a, 1969b).

Large Aerosols. On the basis of the direct microchemical analyses that are possible in this size range, much more detailed information can be gained concerning the chemistry of the "large" and "giant" nuclei. Analyses to date show great differences in

---

(Footnote 18 continued)

and reevaporation study of individual particles Figure 19) may be of importance here. If this cycle is observed under exact thermodynamic conditions in arbitrary particle mixtures, we find surprising deviations between the behavior of the platinum nuclei used in Figure 19 and other particles. Two photographs of this type have been reproduced in Figure 30.

The experimental conditions were the same as those described for Figure 19. The original Pt-aerosol contained room aerosol with cigarette smoke as an admixture. In addition to uniform tracks with partially visible ellipses, which must undoubtedly be interpreted as Pt-aerosol particles, we recognize those which have a clearly different behavior, are highly deformed, and remain partly visible during the entire cycle. Some helical tracks are also particularly remarkable.

Without being able to offer an interpretation of this behavior, it can be said that recording of the cycles by this method, with systematic variation of the aerosol sources and the secondary chemical conditions, allows us to expect information concerning the particle species and their composition.

chemical composition of this aerosol component from one monitoring station to the next.

In his studies in Frankfurt/Main and its vicinity, and at Round Hill on the U.S. east coast (about 80 km south of Boston, Mass.), C. E. Junge (1953, 1954, 1956) analyzed the two size ranges separately and found characteristic differences in composition: in the large nuclei—between 0.08 and 0.8 μm in size—NH_4^+ and SO_4^{--} ions predominate. The concentration ratio of the two constituents indicates a high $(NH_4)_2SO_4$ content of the nuclear material. In the giant nuclei—0.8-8 μm size—a high NaCl concentration becomes preponderant (see Table VIa).

Table VIa

Mean Values of Some Cations and Anions in Soluble Material of Large Nuclei and Giant Nuclei*

	Mean Values in $\mu g/m^3$	
Ionic Species	*Large Nuclei*	*Giant Nuclei*
Cl^-	0.03	1.22
Na^+	0.07	1.15
SO_4^{--}	4.58	1.16
NH_4^+	0.83	0.19
NO_3^-	0.06	0.69

*According to measurements of C. E. Junge (1954) at Round Hill (mean of 27 single diurnal analyses). Data in $\mu g/m^3$ of air. Large nuclei: radii between $\sim$ 0.08 and 0.8 μm; giant nuclei: radii between $\sim$ 0.8 and 8 μm.

The change in composition in the transition from one size range to the next at a location near the coast compared to the data in Figure 28 suggests that the aerosol involved, which is evidently a mixture of components of continental origin with those of marine origin, shows a different mixing ratio in the two size ranges insofar as the continental (marine) component predominates in the range of large (giant) nuclei. Interesting information is obtained from a comparison of aerosol analyses of this type as a function of the type of climate as shown in Figures 31 and 32.

While the mean values of SO_4^{--}, NH_4^+ and NO_3^- ions decrease similarly in the two types of nuclei in the transition from a

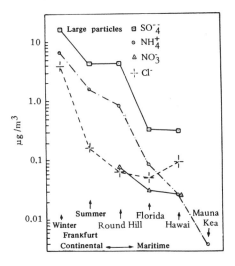

Figure 31. Mean values of aerosol analyses in the range of large nuclei as a function of climate according to C. E. Junge (1956).

continental to a maritime climate, the Cl⁻ ions show a different be- havior: in the large nuclei, their concentration first decreases, then increases again in the purely maritime region, while giant nuclei exhibit a more general increase of Cl⁻ ions.

This comparison confirms the above-mentioned hypothesis that the marine constituent is manifested more in the giant nuclei than in the large nuclei as indicated by the Cl⁻ content. Without dwelling on this further, it can be deduced from the above that a

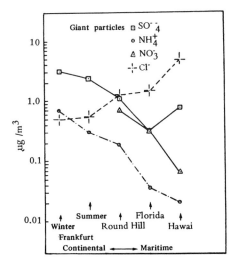

Figure 32. As Figure 31 for the range of giant nuclei.

separation of marine and continental constituents in an actual aerosol is basically possible. The Cl⁻ and NaCl fraction in the continental aerosol plays a special role here and should be briefly discussed once more.

As shown by Figure 28 and confirmed again here, the Cl⁻ ion content over the continent remote from the coast can no longer be of marine origin. Figure 8 of Chapter 2, which showed a surprisingly high Cl_2 concentration and thus production over the continent, points in the same direction. Two possibilities exist to explain this finding (in addition to any direct Cl_2 production by air-polluting processes): decomposition of oceanic NaCl and direct penetration of NaCl into the atmosphere by turbulent transport of soil material.

According to R. C. Robbins, R. D. Cadle and D. L. Eckhardt (1959), the first possibility can be interpreted by reactions of the type represented by Equation (29) in the presence of NO_2 and H_2O under the influence of radiation:

$$NaCl + 2 NO_2 = NaNO_3 + NOCl$$
$$NOCl + H_2O = HCl + HNO_2 \qquad (29)$$
$$NOCl + h\nu = NO + Cl$$

Such a Cl loss in marine aerosol entering the continent is qualitatively confirmed by the measurements of rain water showing a decrease of the Cl^-/Na^+ ratio from about 1.8 over the ocean and on the coast to values of 1 and less in the continental interior.

The second possibility is supported by the increase of the K^+/Na^+ ratio from a value of about 1/20 on the coast to about 1 in the continental interior (potassium and sodium are found with approximately equal frequency in rock material). The same direction is indicated by the Ca^{++} ions found in rain water in a ratio of K^+ and Na^+ of about 10:1, corresponding to the rock mineral composition (C. E. Junge and R. T. Werby, 1958). This suggests a considerable proportion of soil material in continental aerosols.

In Greenland ice the ratio of these three cations is also considerably different than in sea salt (C. E. Junge, 1963); on one hand, this indicates a considerable proportion of soil substance of continental origin even in the aerosol of this very clean atmosphere, and on the other hand it suggests the probability that such material must also be present in small aerosols that survive long transport paths. The latter is confirmed by observations of Sahara dust over the Atlantic west of North Africa, which can be transported through the straits up to Florida.

Series analyses conducted by the U.S. Public Health Service show the presence of a number of metal compounds in urban aerosols, which can be explained by air pollution processes (U.S. Dept. of Health, Education and Welfare, 1969). Table VII lists some of these results together with the measured sulfate, nitrate and ammonium values. The finding that the metropolitan aerosol with a mean total weight of 105 $\mu g/m^3$ contained a mean of 6.8 $\mu g/m^3$ of benzene-soluble, *i.e.,* organic material, is of particular interest.

Table VII

Mean and Maximum Aerosol Concentration in U.S. Cities, 1960-1965*

Material	Mean	Maximum
Iron	1.58	22.00
Lead	0.79	8.60
Manganese	0.10	9.98
Copper	0.09	10.00
Vanadium	0.05	2.20
Titanium	0.04	1.10
Tin	0.02	0.50
Zinc	0.67	58.00
Nickel	0.03	0.46
Arsenic	0.02	**
Beryllium	0.0005	0.010
Cadmium	0.002	0.420
Chromium	0.015	0.33
Molybdenum	0.005	0.78
Antimony	0.001	0.160
Nitrates, NO_3^-	2.6	39.7
Sulfates, SO_4^{--}	10.6	101.2
Ammonium, NH_4^+	1.3	75.5
Benzene-soluble organics	6.8	**
Total aerosol content	105.0	1254

*Data in $\mu g/m^3$. (U.S. Dept. Health, Education and Welfare, 1969).
**No data are available.

The organic components in the aerosol in which hydrocarbons and hydrocarbon derivatives can be detected probably originate from natural and anthropogenic sources. Thus, A. Goetz and O. Preining (1960) have reported on the production of considerable quantities of such substances with a size range of large nuclei in the forests, mountains and desert regions of the western USA. According to A. Goetz (1960), the Los Angeles smog also contains organic compounds, the formation of which is attributed to gas reactions between hydrocarbons and trace gases (O_3, NO_2, SO_2).

Mineral soil material evidently reaches the atmosphere in considerable quantities as a result of strong air motion (dust storms) as can be indirectly concluded from loess deposits and deposits on the ocean floor. Furthermore, direct studies of H. Glawion (1938) in Arosa and of C. E. Junge in Florida demonstrate the presence and long-distance transport of Sahara dust. In addition, large amounts of aerosols are introduced into the atmosphere by volcanic activity (for example, see *NACR Quarterly,* February 1969).

Stratospheric Aerosols of Terrestrial and Extraterrestrial Origin

According to aerosol studies in the stratosphere that have been extended up to an altitude of 27 km thus far (C. E. Junge, 1961, and C. W. Chagnon and C. E. Junge, 1961; see also Figures 35 and 36), aerosols reach these altitudes in considerable quantity even from lower layers. A marked difference was observed in the behavior of the Aitken nuclei and large nuclei measured in these studies: while the former decrease to less than 1% from the lower boundary of the stratosphere up to 20 km altitude according to Figure 36, the number of large nuclei increases, reaching a maximum at an altitude between about 16 and 23 km (see Figure 37). This phenomenon leads to the conclusion that the Aitken nuclei clearly penetrate the stratosphere from below and, in the interaction of vertical transport and sedimentation, they form the altitude profile determined in the measurements.

In contrast, the altitude distribution of large nuclei indicates a particle current in a downward direction from the maximum zone. However, since the chemical composition of these particles reveals their terrestrial origin—98% of the total weight is represented by SO_4^{--} and 10% by Al, Si, K and Ca (C. E. Junge and J. E. Manson, 1961)—it may be concluded that they are probably formed in the stratosphere directly by oxidation of SO_2 and H_2S. This is also supported by the finding that the maximum zone of these particles shown in Figure 36 coincides approximately with the maximum of the ozone layer (see Figure 9).

Very little is known about aerosols of extraterrestrial origin. They are difficult to identify beside the aerosols of terrestrial origin, which are also present in much larger quantity in the stratosphere.

Particles of cosmic origin, to the extent to which these have been definitely identified in the atmosphere, consist of small magnetizable Fe_2O_3- or FeO-sphers with radii of about 3-100 μm (see, among others, W. D. Crozier, 1960, P. W. Hodge, 1961, and P. W. Hodge and F. W. Wright, 1962). In part, they might represent residues of molten meteoric material, but a larger part originates perhaps from the nebula of the zodiacal light to the extent to which this penetrates the atmosphere, reaching melting temperatures in the process. According to E. J. Oepik (1957) the size of the zodiacal light particles ranges between a few tenths of a μm and 300 μm, with a distribution function of the following character

$$\frac{dn}{dr} = c \cdot r^{-p} \tag{30}$$

with possible values of p = 2.8 and c = 10^{-20}.

The extraterrestrial particles are of no significance in the terrestrial aerosol balance.

METEOROLOGY AND CLIMATOLOGY OF AEROSOLS

The discussion of aerosols to this point has given primary attention to their formation and physical as well as chemical properties, while the problems of their dispersion, distribution and the abundance of their sources, their life history in the atmospheric space, their significance as a part of the atmosphere and other pertinent questions could only be mentioned briefly. These neglected problems will therefore be considered now.

The dispersion of aerosols of natural and anthropogenic origin in the atmosphere is determined by the interaction of the distribution and abundance of the sources, atmospheric transport mechanisms and redeposition. The resulting aerosol cycles are influenced by geography, climatology, meteorology and aerology. In addition, depending on external conditions, dynamic and chemical changes of the aerosols can also have a concomitant structuring effect.

The test data available for an analysis of aerosol cycles differ quite considerably for different particle size ranges. The most complete data exist for the range of the Aitken condensation nuclei. In the range of large nuclei and particularly giant nuclei, the data base is much smaller. However, these differences can be compensated for to a certain degree by applying the Junge law of aerosol size distribution, since this law, when carefully applied, allows the extrapolation of data from one size range to the others.

As far as the trend of aerosol cycles with time is concerned, we should note that information on the subject can be obtained from atmospheric radioactivity (for details, see Chapter 4). It can be ascertained that the mean residence times of the aerosols in the troposphere fall into the order of days to weeks, while in the lower stratosphere they are in the order of months.

Climatology and Geographical Distribution of Aerosols and Their Sources

For a general orientation on the climatological and geographical distribution of aerosols and their sources, we can use a summary of the number of Aitken nuclei in different regions of H. Landsberg (1938) as shown in Table VIII. This summary confirms the hypothesis advanced earlier in this chapter that the Aitken particles are essentially of continental origin. Furthermore, we recognize the clear relationship between the particle counts and population density: the main producers are large cities. Thus, we have confirmation of the finding that measurements of the Aitken numbers can be used as test values for an assessment of air pollution.

Figure 33, which summarizes the results of Table VIII in its most important characteristics, can thus serve as a

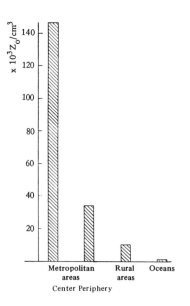

Figure 33. Outline of the distribution of the Aitken nuclei content relative to population density.

Table VIII

Summary of the Mean Count and Range of Variation of Aitken Particles in Different Observation Regions*

Observation Zone	Aitken-particles/cm³		
	Mean	Mean Maximum	Mean Minimum
Large cities	147000	379000	49100
Smaller towns	34300	114000	5900
Rural areas, inland	9500	66500	1050
Rural areas, coast	9500	33400	1560
Mountain stations, 500-1000 m	6000	36000	1390
1000-2000 m	2130	9830	450
Over 2000 m	950	5300	160
Islands	9200	43600	460
Ocean**	940	4680	840

*Compiled by H. Landsberg (1938) on the basis of available data from a total of 171 monitoring stations.

**The values listed for the oceanic atmosphere are probably too high, since numerous measurements performed near the coast are contained in these mean values. According to more recent studies, a mean value of about 100-300 per cm^3 is more probable at a sufficient distance from the coast. In some cases, for example, over Greenland (R. W. Fenn, 1960), the data may still be considerably lower.

qualitative orientation for the general aerosol character at least to the extent to which the latter depends on population density.

Further information concerning climate-determined variations of the aerosol state can be derived from its dependence on the origin of air masses. In its climatological and geographical distribution, the produced aerosol is transmitted from the ground to the air located or moving over a given region, *i.e.*, it produces a certain labeling of the respective air mass that may even be retained over longer distances as shown by observation. Figure 34 shows an example of this relationship derived from a diagram in the same study of H. Landsberg (1938).

Corresponding studies elsewhere led to similar patterns. The fact that the dependence of the particle count on the origin of the air differs from one location to another is not surprising. In its basic principles—lower particle counts in air masses of polar or

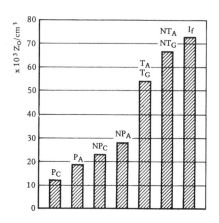

Figure 34. Mean atmospheric content of Aitken particles (Z_o/cm^3) as a function of the origin of the air at State College, Pa., USA according to H. Landsberg (1938).

P_C = Canadian polar air
P_A = Atlantic polar air
T_A = Atlantic tropical air
T_G = tropical air (originating from the Gulf of Mexico
N = "aged" air of respective origin
I_f = indifferent air with subsidence.

maritime origin compared to tropical or continental air masses— the picture in any case remains the same.

Changes in air masses at a monitoring station can exhibit not only modifications in counts but also in size spectrum of the aerosols. Such a case is illustrated in Figure 35, which shows the number and size spectrum of medium and large ions in the case of a cold front in the Hochtaunus Mountains. The ionic spectra in the lower inset, differentiated from top to bottom by shading, are

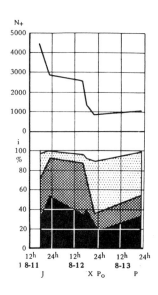

Figure 35. Changes in the number (top) and size spectrum (bottom) of medium and large ions in a cold-air front in the Hochtaunus Mountains (according to N. Weger, 1934).

I = "indifferent" air
X = mixed air
P_o and P = polar air
(On the representation of the ionic spectrum, see explanations in text).

summarized in the usual four groups of ultra-large ions, Langevin ions, large medium ions and small medium ions.[19] Thus, with a transition to polar air with a lower aerosol content, we find a marked shift of the spectral peak toward smaller dimensions.[20]

Altitude Distribution of Aerosols

The mean values for the Aitken particle counts for mountain stations listed in Table VIII suggest a rapid decrease of the aerosol content with altitude. This has been confirmed by numerous field studies for the region of the lower troposphere, which showed an approximately exponential decrease of the mean particle counts with altitude. Figure 36 shows the Aitken particle counts up to 6 km altitude as a profile in curve *a* measured during the climbing

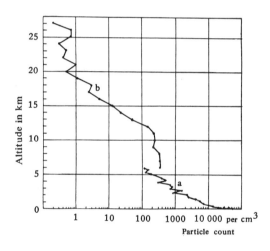

Figure 36. Mean altitude profiles of condensation nucleus counts.
Curve *a:* Mean number of Aitken particles according to 12 aircraft flights of H. Weickmann (1955).
Curve *b:* Mean Aitken particle count between 6 and 27 km altitude after climbing phases of 7 aircraft flights by C. E. Junge (1961).

phase of 12 aircraft flights of H. Weickmann (1955). The picture changes in the upper troposphere: The decrease with altitude becomes noticeably slower, usually already starting at 4-5 km, and finally changes into a more or less constant character with altitude almost up to the tropopause. Curve *b* shows the mean altitude

---

[19] Dimensions of the respective ions according to H. Israel (1957):
Small medium ions: Diameter between 0.00132 and 0.0156 μm
Large medium ions: 0.0156-0.05 μm
Langevin ions: 0.05-0.124 μm
Ultralarge ions: 0.124 to about 0.2 μm
[20] With regard to additional experiences concerning the problem of "air mass or air parcel and ionic spectrum," see N. Weger (1934).

trend of Aitken particles from 6 to 27 km altitude measured in the climbing phase of 7 aircraft flights of Junge (1961). In the range of altitude-independence, the Aitken particle count in the upper troposphere ranged between about 60 and 600 per cm^3 with a mean of about 300/cm^3.

Aitken nuclei and large nuclei show the same behavior with regard to their altitude distribution in the troposphere, while they differ in the stratosphere: while a decrease of Aitken particles to less than 1% can be observed from the lower boundary of the stratosphere up to 25 km altitude according to curve b of Figure 36, available data for the large nuclei surprisingly show an increase starting with the tropopause and leading to a maximum at between 16 and 23 km altitude (see Figure 37).

These findings suggest the following explanations and conclusions. On the basis of the behavior of Aitken nuclei in the troposphere of the continents—decrease to about 1/100 of the ground-level value up to the middle troposphere, followed by transition to a practically altitude-independent concentration up to the troposphere—it can be assumed that the Aitken aerosol forming on the surface is subject to certain modifications during its upward transport mediated by turbulent mass exchange, and it reaches a stable state with a highly reduced count only after some time. The factors that may cause these changes consist of coagulation[21] in cloudless weather without precipitation[22] to which so-called rainout (incorporation of particles into the cloud and precipitation components) and so-called washout (by falling precipitation) must be added in the presence of clouds and precipitation. It may be assumed that up to about 3 km altitude the reduction in particle count by coagulation predominates over that due to rainout in the lower layers, but that the ratio of the two effects is then reversed.

The finding that the mean number of Aitken particles in the upper continental troposphere is approximately identical to that found in the lower layers of the marine atmosphere is of particular interest. A possible explanation of this phenomenon would be that the amount of a few hundred Aitken nuclei is considered as the fraction that is uniformly admixed to the total atmosphere as the Aitken background aerosol. In accordance with this concept, the lower continental troposphere would then be the source

---

[21] The measurements are practically all "good-weather recordings."

[22] Sedimentation can be practically ruled out, since Aitken particles and large particles show the same behavior in this zone of the atmosphere.

region in which this fraction of continental aerosol is formed and transformed into the Aitken background aerosol. An application of this theory to the field of large nuclei seems indicated in view of their similar behavior in the continental troposphere, but this has not yet been confirmed by observation.

Reference has already been made (page 68) to the difference in behavior of the Aitken nuclei and the large nuclei with regard to their altitude distribution in the stratosphere and the conclusions to be derived from this. The transition of the altitude distribution of Aitken nuclei from practically altitude-independent concentrations or to those that decrease only very slowly with altitude in the upper troposphere to a rapid altitude decrease in the stratosphere can be explained easily by the decrease of the eddy diffusion coefficient by 1-2 orders of magnitude in the transition from the troposphere to the stratosphere. Moreover, the altitude decrease clearly allows the conclusion that these particles are of tropospheric origin and reach the stratosphere by mixing processes.

The completely different behavior of large nuclei with their rise to a maximum at altitudes of about 20 km, as mentioned earlier, must mean that these particles are formed directly in the stratosphere. Since their altitude distribution shows similarities with that of the ozone layer, a causal relationship can be assumed and particle formation by oxidation of sulfurous gases, H_2S and SO_2 into SO_3, with the formation of sulfuric acid or of sulfates can be suspected. The presence of a haze layer of the type corresponding to Figure 37 at altitudes of about 20 km is well-known and evidently represents a worldwide phenomenon as is also apparent from atmospheric electrical and optical studies (see, for example, O. H. Gish and K. L. Sherman, 1936, E. G. Gibb, 1956, F. Volz, 1961, and others). It is probable that there is a relationship with the cirrus cloud forms that are observed occasionally at 25 km altitude.

Aerosols and Weather

The relationships mentioned thus far are subject to extensive differentiation by meteorological influences. As expected, the aerosol state of the air is closely related to weather phenomena. This can be demonstrated by statistical correlations of the aerosol content with individual meteorological elements as well as by the graphic representation of single examples of particularly distinctive relationships.

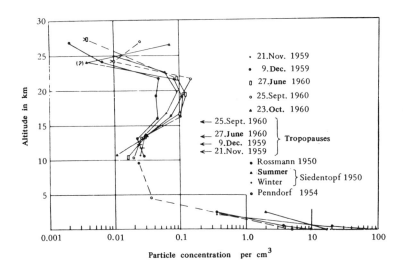

Figure 37. Five stratospheric altitude profiles of large nuclei (mean radius 0.15 μm) according to aircraft sampling by C. W. Chagnon and C. E. Junge, 1961. At the bottom right, older measurements in the lower troposphere by other authors are shown for a comparison.

Particularly in the past, the former method has been used and the causal connection between weather and aerosol was sought in statistical relations. For example, H. Landsberg (1938), using the entire observational data available at the time, investigated the dependence of the Aitken particle count on air mass, wind direction and intensity, visibility and humidity. He arrived at the result that the relations with the individual elements are not as unambiguous as would be expected. While the relations with the origin of air masses and their displacement distances as well as with the wind direction are qualitatively clear, the picture becomes obscure and contradictory with regard to the other elements. With increasing wind velocity, the mean number of nuclei decreased more or less prominently in 18 of 24 test series from different monitoring stations. In five cases, the inverse situation was found, while no reaction was observed in one. With regard to visibility, a decrease in number of particles with increasing visibility was found in 10 of 17 measuring series, while the inverse relation existed in 3 cases and there was no clear dependence in 4. The relation of the particle counts to relative humidity is particularly inconsistent. Of the 14 available series on this aspect, the data originating from large cities showed a trend of increasing nuclei counts with relative humidities ranging from 70 to 100%, while a decrease in particle count appeared to predominate in the remaining cases for the same humidity range. A trend toward a decrease is generally observed between 30 and 70% R.H.

It is difficult to explain the different patterns of particle counts relative to the meteorological elements at different monitoring stations in individual cases without more detailed special studies. In general, the reasons for the differences probably must be sought in local conditions. Furthermore, it must be kept in mind that the value of such statistical correlations with one element or another in principle is limited by the fact that all other meteorological influences are also present and active in each of them. From this standpoint, a study of particularly outstanding examples may offer more succinct and reliable information than statistical analysis. Otherwise, the picture generally becomes clearer if one starts inversely from the meteorological event and its laws, then derives the anticipated aerosol reactions on their basis and tests these empirically.

Periodic Variations. Like all atmospheric parameters and properties, the aerosol content is subject to typical variations in the course of a day and year. If we again consider the Aitken nuclei to be representative, the following examples can be cited.

Figures 38-40 show individual results from a one-year monitoring series of Aitken counts in Payerne, Switzerland (P. Ackermann, 1954). During the winter months, the particle counts show a simple periodic variation in the course of the day, with a minimum during the late night hours and a maximum in the afternoon to early evening. In contrast, in the summer months, two maxima and two minima are clearly evident in the diurnal vairations. The annual fluctuation exhibits a simple periodicity with minimum (maximum) values in the winter (summer) months in Payerne.

A significantly different picture is obtained from the particle count recorded at the high mountain station of Jungfraujoch-Sphinx at approximately 3600 m altitude (H. Israel, 1957a; E. Herpertz, 1957): With practically no annual variation, the diurnal trend shows a simple periodicity over the entire year (see Figure 41).

Together with studies from other stations (see, for example, F. Verzar, 1967 and others), the following three types of diurnal

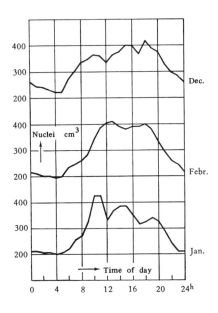

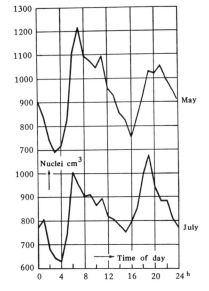

Figure 38   Figure 39

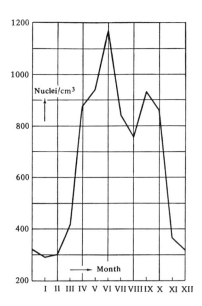

Figure 38. Mean diurnal varia-
tion of the condensation nuclei
counts during the winter months of
December, January, February in
Payerne, Switzerland, according to
P. Ackermann (1954).

Figure 39. The same for the
summer months of May and July.

Figure 40. Annual variation of
condensation nuclei counts in Payerne, Switzerland, according to P.
Ackermann (1954).

Figure 40

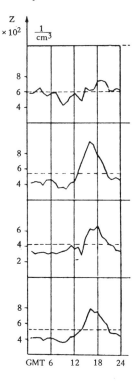

Figure 41. Mean diurnal variation of the condensation nuclei counts at the Jungfraujoch-Sphinx high-mountain station according to E. Herpertz (1957). From top to bottom, the variations during the seasons of "winter" (November-February), "equinoxes" (March-April and September-October) and "summer" (May-August) and during the entire year.

and annual variations can be distinguished in the behavior of the particle counts if local conditions are neglected:

Group 1: Simple periodic diurnal variation during the entire year.

Group 2: Simple-periodic diurnal variation in winter and double-periodic in summer.

Group 3: Double-periodic diurnal variation for the entire year.

In Group 1, the mean particle counts are generally low, while Group 2 has medium-high and Group 3 high particle counts. The annual variations show a different situation: their amplitudes of fluctuation seem to increase from Group 1 to Group 3. In regions with an abundance of nuclei and high nucleus production, the extreme values usually are the inverse of the trend shown in Figure 38.

The behavior of the aerosol in its regular diurnal and annual variations corresponds exactly to expectations for a property or

an admixture originating from the surface of the earth being trans-
ferred to the atmosphere under the influence of atmospheric mass
exchange. Consequently, we find close parallels with the behavior
of other elements, for example, the diurnal variations of the water
vapor content or, even closer, with the diurnal and annual periodicity of the electrical field in the atmosphere (see, for example,
H. Israel, 1961). The parallel in the case of the water vapor content is purely correlative, since both are subject to a similar influence, while in the case of the parallel with the electrical field of
the air, a causal relationship exists to the extent that in magnitude
and variation, this field is substantially codetermined by aerosols
and their behavior (see page 83).

Individual Effects. Within the framework of these general relationships, the effects of weather on the aerosol state of the air can
be documented by numerous single examples.

The relationships to wind direction, which reflect the influence
of the location and abundance of aerosol sources on the air reaching the sampling station (see Figure 33), are particularly marked in
a large number of variants. In urban zones, clear relationships can
always be recognized with the distance covered over the urban region and with particularly abundant industrial interference zones.
Outside of the cities, the particle count varies to a considerable degree depending on whether air motion allows the haze cover of
the city to advance to the sampling station or not. The same group
of phenomena includes, for example, aerosol reactions to changes
of continental and sea winds along the coast. (Examples with numerical data concerning fluctuations of the Aitken values can be
found in the survey of H. Landsberg, 1938.)

With regard to this dependence on air masses, it should still be
added that these are generally weaker in regions of high rather
than in those of low population density. This can be explained
easily by the fact that these differences are evidently increasingly
masked by impregnation of the air with high urban particle counts.
On the other hand, it indicates that when the air moves over zones
of high aerosol production, it changes its aerosol character relatively quickly, so that the latter reflects the path of the respective
air mass more than its origin. The same is known for the aerosol
variations connected with air mass exchanges. These, too, are
more pronounced the lower the mean aerosol level is of a given
monitoring station (see L. Schulz, 1933). The relationship of the
aerosol state of the air with its previous history extends from

that with the origin of air masses via the dependence on wind direction up to small local influences in connection with the directional nature of wind gusts in turbulent flow. It becomes marked in the aerosol differences of minute air parcels of turbulence cells that can produce a type of aerosol turbulence, according to Figure 42. Experience has shown that the range of variation again is greater outside of but near an urban zone than within the urban zone itself.

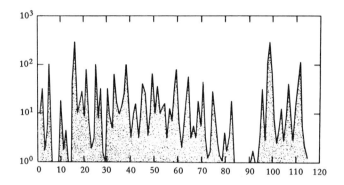

Figure 42.   Example of "aerosol turbulence." Abscissa: time in minutes. Ordinate: aerosol concentration in arbitrary units (from A. C. Stern, Air Pollution, 1968).

Another controlling influence of aerosol characteristics resides in the effect of turbulent exchange: if s is a property of the air, for example, its aerosol content, then generalization of the second Fick diffusion law, limited to the altitude dependence of the phenomena, leads to the following relation:

$$\frac{\partial s}{\partial t} = K \frac{\partial^2 s}{\partial h^2} + \frac{\partial K}{\partial h} \frac{\partial s}{\partial h} \qquad (31)$$

K = eddy diffusion coefficient
h = altitude

which relates the variation of s with time with the variation of this property with alitutde and takes into account that the eddy diffusion coefficient K is no longer spatially constant but is altitude-dependent in contrast to the diffusion coefficient in molecular kinetics. The product K · ρ (ρ = air density) is known as the exchange coefficient A. Increasing from small values at ground level, the values have an order of magnitude of about 50-100 [g cm^{-1} sec^{-1}] in the atmosphere compared to values of about 10^{-4} [g cm^{-1} sec^{-1}] for the corresponding coefficient in molecular processes.

The exchange is closely connected with atmospheric temperature stratifications: with a stable stratification, its influence is greatly reduced, while under unstable conditions, it is increased. Since this temperature stratification passes through regular diurnal and annual fluctuations and is also directly related with weather development, characteristic variations result for the turbulent exchange and its change with altitude, which are reflected in corresponding reactions of aerosol conditions, thus furnishing an explanation for their periodic and aperiodic variations.

Figure 43 shows the diurnal variations of the exchange coefficient A for different altitudes above ground level (H. Lettau, 1941). If the upward transport under the influence of exchange now becomes greater than the resupply of a given element from the bottom, this leads to a corresponding depletion in lower atmospheric layers. In the summer trend of the diurnal curves for the condensation nucleus content this evidently becomes apparent by the depression of the curve at midday as indicated by the variation patterns shown in Figures 38-40 and described here.

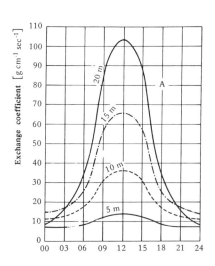

Another example demonstrates the influence of temperature stratification on the exchange-determined altitude distribution of the particle counts: we know from experience that the vertical temperature variation in the atmosphere is not uniform, but is interrupted by so-called inversions in which the temperature decrease with altitude is interrupted by temperature constancy or increases. Such layers reduce or interrupt vertical exchange. This must then become

Figure 43. Diurnal variation of the exchange coefficient at different altitudes (according to H. Lettau, 1941).

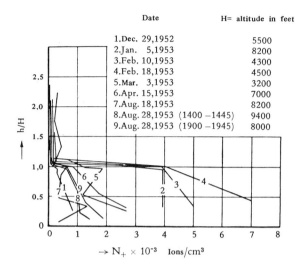

Date		H= altitude in feet
1.Dec. 29,1952		5500
2.Jan. 5,1953		8200
3.Feb. 10,1953		4300
4.Feb. 18,1953		4500
5.Mar. 3,1953		3200
6.Apr. 15,1953		7000
7.Aug. 18,1953		8200
8.Aug. 28,1953	(1400 −1445)	9400
9.Aug. 28,1953	(1900 −1945)	8000

$\rightarrow N_+ \times 10^{-3}$ Ions/cm³

Figure 44. Altitude distribution of the "charged" suspensions in nine aircraft flights according to R. Sagalyn and G. A. Faucher (1954). The altitude criterion selected is the ratio h/H between actual altitude and altitude H in which the sudden decrease occurs (inversion altitude).

evident in sudden changes (decreases) of the aerosol content as has been observed in aircraft flights. Figure 44 shows this on the basis of the aircraft tests of R. Sagalyn and G. A. Faucher (1954).

Aerosols as Mediators of Meteorological Influences on the Atmospheric Electrical Phenomena. As we know, atmospheric electrical phenomena are closely connected with meteorological events. If we neglect for the moment the effects connected with the atmospheric water cycle—cloud formation, precipitation phenomena—in which special types of laws are involved, the relationships of meteorology and atmospheric electricity are causally connected with the presence of suspensions in the air and their variations.

The mechanism of this involves the interaction of aerosol production and transport under the influence of meteorological events with the recombination processes discussed earlier (see Chapter 2, page 13) that modify the charge and conductivity conditions in the atmosphere.

For a quantitative consideration of these relationships, we require a functional connection between the specific electrical resistance w of the air with the ionization strength q and the suspension content N of the air. The form of the functions

$$w = f(q, N) \qquad (32)$$

or, if q is considered to be constant in time in first approximation,

$$w = f'(N) \cdot \text{const.} \qquad (33)$$

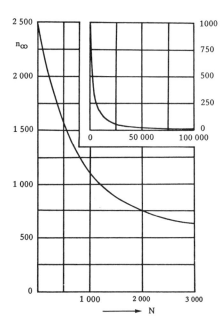

is determined by recombination processes. To illustrate this by an example, Figure 45 shows the equilibrium small ion counts n_∞ as a function of the particle count N which can be calculated by integration of Equation (8) with plausible assumptions concerning the recombination coefficients, ionization strengths q and relation between charged and uncharged nuclei (for further details, see H. Israel, 1955/56).

The electrical elements of the air, *i.e.*, field strength E, current density i and conductivity Λ, are interrelated by the Ohm law

$$E \Lambda = i \qquad (34)$$

where the conductivity is given by

Figure 45. Equilibrium small ion count n_∞ for condensation nucleus counts of between 0 and 10^5 per cm^3 calculated by integration of Equation (8) with the following assumptions:

$q = 10$ ion pairs per cm^3 and sec
$a = 1.6 \cdot 10^{-6}$ $[cm^3 \ sec^{-1}]$
$\eta_0 = \eta_c = 10^{-5}$ $[cm^3 \ sec^{-1}]$
$N^+ = N^- = 0.4 \ N.$

$$\Lambda = \Lambda^+ + \Lambda^- = e\,n^+ k^+ + e\,n^- k^- = \frac{1}{w} \qquad (35)$$

Λ^+ and Λ^- represent the so-called polar conductivity, e is the elementary charge, n^+ and n^- are the number of positive and negative small ions, and k^+ and k^- are the corresponding mobilities.

If we assume that the electrical field strength and conductivity of the air depend only on the altitude above ground level, *i.e.,* they have no horizontal components—a generally customary and permissible simplification—then the following expressions are valid in stationary equilibrium, which is characterized by altitude independence of i:

$$E \Lambda = \text{const} \tag{36}$$

or

$$E \sim w \tag{37}$$

This explains the similar trend of the electrical field and the suspension (nucleus) content of the air, which also show the same direction in their variations.

Functions (32) and (33) can be obtained by calculation or measurement. In individual particle count ranges they can be represented by simplified approximations. Thus, for the range of smaller particle counts, a logarithmic representation seems appropriate and has been fairly well confirmed by experimental data obtained on the Jungfraujoch-Sphinx (see Figure 46).

Such experimental determinations always contain considerable scatter of single data as a typical secondary phenomenon.

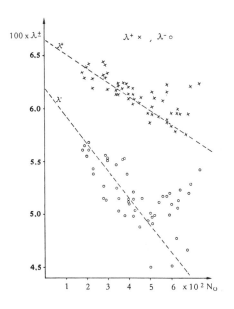

Figure 46. Dependence of the two polar conductivities Λ^+ and Λ^- on the particle count at the Jungfraujoch-Sphinx station according to E. Herpertz (1957). The plotted points are mean values of the range.

This can still be recognized in Figure 46 in spite of their combination into mean values of a range. The reason for this great scatter is the considerable time necessary to reach equilibrium between n or Λ and N, a time span that is generally not available in atmospheric processes. Thus, various degrees of approximation of the final state will be found side-by-side in such experimental data plots. Another example of this is also shown in Figure 47 representing the results of an annual series of measurements of n and N. The figure shows the small ion content n^+ and n^- as a function of $N^{-\frac{1}{2}}$

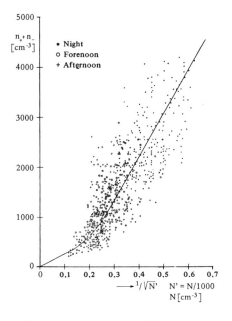

Figure 47. Small ion count $n^+ + n^-$ in relation to the reciprocal root of the particle count N on the basis of a one-year study in Garmisch-Partenkirchen (R. Reiter, W. Carnutz and R. Sladkovic, 1968).
Abscissa values: $(0.001\ N)^{-\frac{1}{2}}$
0.1, 0.2 . . . 0.7 thus correspond to the following N-values:
100,000; 25,000; 11,200; 6,300; 4,000; 2,270; 2,070.

It can be recognized in this case that the range of variation decreases with increasing particle count. This is also a typical phenomenon. It is explained by the fact that the stationary end state is approached more rapidly the greater the value of N, as has been confirmed theoretically (H. Israel, 1957c) as well as experimentally in the laboratory (C. G. Stergis, 1954).

On the basis of these considerations on the mediator role of aerosols in the development of relationships between meteorology and atmospheric electricity, these phenomena can often be easily explained. Thus, the parallelism of the diurnal and annual periodic variations of the atmospheric electrical field with those of the aerosol content becomes readily understandable on the basis of Equation 37. Considerations on exchange and its effect on aerosol distribution largely explain the altitude trend of the atmospheric electrical field and conductivity. The electrical variabilities of the atmosphere are undoubtedly related with aerosol turbulence.

In other phenomena, such as the effect of the sunrise or of the brightness, the mediator role of aerosols perhaps may not be quite as clearly recognizable, but it is nevertheless probable or conceivable.

For details concerning these effects, see H. Israel (1955, 1955/56, 1957a, 1957b, 1961), and others.

Aerosols and Precipitation

Particularly close relationships exist between atmospheric aerosols and the processes of cloud formation and precipitation. As mentioned earlier, a water cycle in the atmosphere as we know it would not be possible in the absence of aerosols. On the other hand, it is precisely this water cycle, made possible by aerosols, that represents the most effective process for their removal from the atmosphere.

The incorporation of aerosols into cloud and precipitation particles takes place by processes of *rainout* and *washout*. The former term combines all processes of aerosol incorporation in cloud and precipitation particles to the extent to which they take place within the clouds, while washout refers to the incorporation of aerosol material by falling precipitation outside of the clouds.

The rainout process is composed of the condensation process and subsequent incorporation of aerosol particles by addition to existing water droplets. Water vapor *condensation* takes place on aerosol particles. Consequently, depending on their chemical composition as dissolved or suspended substance, these particles are incorporated into the cloud and precipitation particles and impose their chemical properties on them. Since the number of cloud or fog droplets as a rule is considerably smaller than the number of existing aerosol particles—at least in the case of continental aerosols—this means that only a part of these are activated, *i.e.*, are utilized as a condensation nucleus. The selection of nuclei that are activated depends on the aerosol properties themselves, especially their size. As we know by experience, it is primarily larger particles, *i.e.*, large nuclei and giant nuclei, that serve as starting points for condensation, while the smaller particles of the Aitken nuclei are activated only to a small degree. As a consequence, the air space between fog or cloud droplets must contain primarily an aerosol consisting of Aitken nuclei; this has been confirmed by particle counts in fog and cloud atmospheres.

In this mixture of droplets and inactivated aerosol particles, a tendency exists for the latter to add to already existing water droplets. This takes place by diffusion and coagulation under the influence of Brownian motion or as a result of the so-called Facy effect. Addition under the influence of Brownian motion can be represented according to Equation 26 as follows:

$$\frac{dn}{dt} = -4 \pi D R N n \qquad (38)$$

where D and n are the diffusion coefficient and number of aerosol particles per cm^3, and R and N are the radius and number of droplets in cm^3.[23] Integration of (38) results in an exponential decrease of n with time, which under otherwise equal conditions is directly related to the diffusion coefficient of the particles (half life inversely proportional to the diffusion coefficient). The tendency of particle decrease with time for $R = 10 \mu$ and $N = 200$ per cm^3, *i.e.*, for the approximate conditions prevailing in cumulus clouds, is characterized by the following half-lives for aerosols of different size:

$$r = 0.01 \ \mu m \qquad \text{half life} = 38.5 \text{ min}$$
$$0.03 \qquad\qquad\qquad 228 \text{ min}$$
$$0.1 \qquad\qquad\qquad 38 \text{ h}$$

In other words, it is a relatively slow process.

The Facy effect is the phenomenon observed for the first time by L. Facy (1955, 1958) that aerosol particles in a water vapor stream are transported to the surface of a droplet that is growing by condensation and away from the surface of an evaporating droplet. This effect is shown in Figure 48. The flux of particles to the condensing droplets can be recognized in the left part of the figure, while the particle-free space around the evaporating drop is apparent in the right. This type of particle transport depends only

---

[23] The simplification compared to Equation (26) results from the fact that the diffusion coefficient of the droplets can be neglected compared to that of the aerosol particles and that their radii are very small compared to those of the droplets.

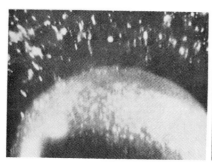 

Figure 48. The "Facy effect." Left: flux of aerosol particles to a condensing droplet of a few μm radius. Right: aerosol-free space surrounding an evaporating droplet of 0.5 μm radius (L. Facy, 1962).

on the water vapor gradient and is independent of the size and shape of the aerosol particles.

With regard to the quantitative contributions of the three effects to the incorporation of aerosols in cloud and precipitation particles, only approximate information can be obtained. Without a doubt, the condensation process is most active. On the basis of a detailed discussion of the condensation process, C. E. Junge (1963) arrived at the conclusion that about 5% of the particles in continental aerosols of high particle counts are "activated" (used for condensation), that this percentage can increase to 50% in a continental aerosol of low particle count, and that a 90-100% activation is probable for marine and tropospheric aerosols with less than 200-300 particles per cm^3. According to this data, the addition process due to Brownian motion is significant practically only in the range of microaerosols.

Opinions differ concerning the influence of the Facy effect. L. Facy (1962) himself comments in this connection that the air space between droplets in a cloud with 100 droplets per cm^3 and 10 μm droplet radius is cleansed of aerosols in about 84 min if only half of this time is assumed to be used for droplet growth. Other studies arrive at smaller effective values (for example, P. Goldsmith, H. J. Delafield and L. C. Cox, 1961).

Although this problem cannot be clarified here, we should note a direct consequence of the "Facy effect." It has been known for a long time that the conductivity of the air is greatly reduced in the interior of fog and clouds: on the average it drops to about one-third of its usual value under identical ionization conditions. At certain times, however, systematic deviations from this behavior occur, which very obviously are related temporally to the formation and resolution of fog. For example, we observe that the conductivity of the air decreases markedly 1-2 h *before* the development of fog and, on the other hand, that clearing of fog can be predicted about 0.5-1.5 h in advance by an increase of conductivity in the fog air (for details and literature, for example, see the survey paper of H. Dolezalek, 1963).

This finding, which is of greatest importance for the prediction of fog as well as its resolution, can be explained as follows by the Facy effect: in the first case (conductivity decrease before fog formation), the small ions formed in the air are increasingly transported to the already growing condensation nuclei together with the advancing stream of water vapor, finding its expression in a decrease of air conductivity. In the second case (increase of conductivity before clearing of fog), the addition of small ions to the droplets is prevented or inhibited due to the already beginning evaporation, thus inducing a rise of conductivity (H. Israel and H. Dolezalek, 1964).[24]

The *washout process* develops by the filtering action of the falling precipitation on the aerosol. Its formation is shown schematically in Figure 49. The aerosol particles can follow the deflecting airstream (thin lines) only to a limited extent because of their inertia. Consequently, a part collides with the droplet and is absorbed by the latter (dashed lines). The magnitude of this part increases with increasing particle size and can be characterized by the ratio $d/2a$ (see Figure 49).[25]

---

[24] Experiences concerning the possibility of embedment of aerosol particles in water droplets by rainout processes seem to indicate that these processes be applied and expanded technically for the elimination of aerosols. A variety of such attempts have been made. Thus, H. Israel and G. W. Israel (1962) discuss a method to facilitate radioactive decontamination by enlarging the carrier aerosols by condensation of supersaturated water vapor to such a degree that they can be collected with customary industrial dust precipitators (cyclones and the like) and eliminated from the air stream. P. Goldsmith and F. G. May have described aerosol separation by collection with stream condensation and subsequent separation on cooled surfaces. For additional details, including quantitative data, concerning this problem, see C. N. Davies, 1966.

[25] All particles of a certain size that are located within the heavy trajectory are collected by the falling drop. The ratio of the diameter of the air cylinder bounded by this trajectory d to the diameter of the falling drop $2a$ is therefore a criterion of the efficiency of the washout process. $d/2a = 0$ means that no particles are captured by the drop, while $d/2a = 1$ means that the drop separates all particles from the air column through which it falls.

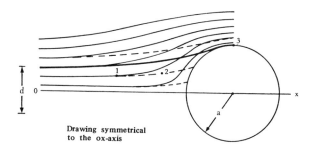

Figure 49. Diagram of the *washout process.*

As a numerical example, we can assume that aerosol particles of 10 μm radius are absorbed by falling drops practically from a space corresponding to the drop cross section ($d = 2a$). With 5 (2) μm particle diameter, this cross section shrinks to about 50 (10)% of the drop cross section (B. J. Mason, 1957). It is difficult to make a sharp distinction between the contribution of rainout and washout to the aerosol decrease.

4

Atmospheric Radioactivity

The third category of trace elements to be discussed covers all of the radioactive materials present in the atmosphere. In contrast to the trace elements considered thus far, a clear distinction into radioactivity from natural sources and artificial radioactivity as the anthropogenic component in the form of nuclear fission products is possible here.

Formally, it would be entirely feasible to classify radioactivity into gaseous and nongaseous trace elements. However, there are various reasons that justify special treatment. We are dealing with quantities at least ten orders of magnitude smaller compared to the trace elements discussed thus far and which are measured in atoms per cubic centimeter, so that a classification according to gases and aerosols frequently becomes meaningless. Table IX illustrates this fact for the incidence of the most important radioactive element in the atmosphere, radon. A few typical mean values have been listed side-by-side for a comparison.

Table IX

Mean Values of the Radon-Content in the Lower Atmosphere in Different Measuring Units

Recording Zone	Radon Content* in $10^{-10} C/cm^3$	Atoms per cm^3	$\mu g/m^3$	ppm
Continent at low altitude above sea level	~ 140	2.5	$9.2 \cdot 10^{-16}$	$7.1 \cdot 10^{-19}$
Mountain region	~ 300	5.3	$19.5 \cdot 10^{-16}$	$18.1 \cdot 10^{-19}$
Coastal region	~ 10	0.17	$0.63 \cdot 10^{-16}$	$0.59 \cdot 10^{-19}$
Oceans and small islands far offshore	~ 1	0.017	$0.063 \cdot 10^{-16}$	$0.059 \cdot 10^{-19}$

*According to H. Israel (1951) and I. H. Blifford, et al. (1956).

Another decisive reason which by itself would justify a separate treatment of radioactive materials consists of their radiation characteristic. The emission of ionizing radiation and the atomic conversions related with this impart new properties to this group of trace elements and, among other things, create the possibility of extremely sensitive methods of identification of these materials.

Essentially, three different processes are responsible for the production of radioactive materials found in the atmosphere: exhalation of radioactive rare gases from the surface of the earth, production of radionuclides by cosmic rays, and production of artificial radionuclides by nuclear weapon tests.

Emanations and their daughter products have their sources in the surface of the earth. The emanations radon (Rn), thoron (Tn) and actinon (An) have a rare gas character and are formed in the earth as intermediate members of the radioactive decay series of the uranium isotopes ^{238}U, ^{235}U and of thorium ^{232}Th. Provided they are produced near the surface, they reach the atmosphere by diffusion and participate in the atmospheric transport process. Their decay in the atmosphere yields a number of radioactive daughter products that are finally converted into a stable lead isotope.

The nucleon component of cosmic radiation in its collisions with atomic nuclei of atmospheric components (essentially nitrogen, oxygen and argon), continuously produces a series of radionuclides. These substances, among which radioactive carbon ^{14}C and hydrogen (tritium) ^{3}H are the most important, are formed primarily in the high atmosphere.

Recently, these two groups of continuously present natural radioactivity in the atmosphere have been joined by artificially produced elements originating from uranium fission and nuclear fusion in the explosion of nuclear and hydrogen weapons. The contribution of nuclear reactors can be neglected, since the escape or radionuclides into the environment has been greatly suppressed by safety measures.

Most of these radionuclides are not gaseous under normal conditions. To the extent to which they are present in atomic form during their production, they probably are rapidly converted into molecular complexes by adding water, oxygen and perhaps trace gases similar to the formation of small ions (see Chapter 3, page 32). Within a few minutes, these primary particles then attach to the aerosols present in the atmosphere as a result of coagulation, so that their physical properties become closely connected with the aerosol cycle. Radioactive decay leads to a continuous reduction of the elements introduced into the atmosphere. Since the lifetime of each element is characterized by a specific half life, a study of the behavior of atmospheric radioactivity in terms of time and place offers a unique possibility for studying dynamic processes in the atmosphere.

For a better understanding of atmospheric radioactivity, the following chapters first discuss radioactive decay and the processes of production of natural and artificial radioactivity. This is followed by a discussion of the dissemination of radioactive materials into the atmosphere and their possible applications for a study of exchange and transport processes.

Radioactive Decay

Elements that emit radiation spontaneously without a supply of external energy are called radioactive. The nuclei of the disintegrating atoms emit either a-rays (helium nuclei with two positive elementary charges) or β-rays (electrons) in this process. As a rule, emission is accompanied by γ-radiation (electromagnetic radiation).

These particles escape from the nucleus at velocities comparable to the speed of light and are gradually slowed down by collisions with air molecules. They lose their velocity and energy essentially by excitation and ionization of the atoms through which they travel. This process results in the formation of atmospheric small ions, among other things (see Chapter 3, page 32).

Radioactive decay takes place according to a statistical law that states that the number of atoms that disintegrate in a given unit of time is proportional to the total number of atoms present. Mathematically the decay law for a given element can be written as follows:

$$\frac{dN}{dt} = -\lambda N \qquad (39a)$$

where N is the number of atoms present at time t and dN are the atoms disintegrating in time dt. The decay constant λ and the mean life resulting from it $\tau = 1/\lambda$ are a constant characteristic for the decaying elements. When (39a) is integrated, we obtain the quantity of radioactive material remaining after time t:

$$N = N_0 e^{-\lambda t} = N_0 e^{-t/\tau} \qquad (39b)$$

where N_O is the number of atoms at the time when decay starts. According to Equation (39b), the quantity of radioactive material decreases exponentially with time. The mean life τ indicates the time in which the number of atoms has decayed to the eth fraction (36.8%) of the original number. Another frequently used unit is the half life T_H in which the number of initially present atoms decreases to one-half. It is calculated from the decay constant as $T_H = 0.693 \cdot 1/\lambda = 0.693\ \tau$. The half lives for radionuclides may extend from fractions of a second up to many billions of years (for example, see Figure 50).

If a substance disintegrates into a series of radioactive intermediates, we speak of a decay series. If we assume that the concentration of initial material is c_1 and its decay constant is λ_1, and that the corresponding notations for the ith intermediate are c_i and λ_i, we obtain the change in concentration of the ith intermediate member with time:

$$\frac{dc_i}{dt} = \lambda_{i-1}\,c_{i-1} - \lambda_i\,c_i\;;\quad i = 1, 2, 3 \ldots \tag{40}$$

where $c_O = 0$. Radioactive equilibrium is present in the decay series when resupply is equal to decay, *i.e.*, $\dfrac{dc}{dt}\,i = 0$. Thus, Equation (40) leads to the well-known relation for radioactive equilibrium

$$\lambda_i\,c_i = \lambda_1\,c_1, \tag{41}$$

which indicates that each member of the decay series has the same activity.

The activity (radioactivity) of a radionuclide is generally expressed by the number of particles emitted per unit of time. The Curie, abbreviated c, is the most common unit and corresponds to the quantity of a radioactive nuclide which produces $3 \cdot 7 \cdot 10^{10}$ disintegrations per second.

Formation of Atmospheric Radioactivity

Contribution of Natural Radioactive Nuclides. The rock crust of the earth contains the long-lived radioactive parent nuclides ^{238}U, ^{235}U and ^{232}Th, which decay radioactively through a series of intermediate products into inactive lead isotopes ^{206}Pb, ^{207}Pb

and ^{208}Pb. In the interior of the earth, these conversion processes take place as a rule *in situ* in compact rock. In contrast, in the earth's crust, displacements can occur by tectonic and volcanic or water and gas movements, so that radioactive nuclides can be transported as a whole together with their carrier rock or as fractions of a group or individual substances. In the following, we will consider only gas transport leading to atmospheric activation.

The uranium and thorium content of rock varies as a function of the type of rock and its location (tabulations can be found in Landolt-Börnstein, 1952, and J. T. Wilson *et al.,* 1956), although the approximate values listed in Table X can be assumed to represent the average in the rock of the continental crust. Accordingly, the uranium and thorium concentrations amount to a few grams per ton of rock. Radioactive equilibrium exists for the individual decay series in the surface of the earth provided the nuclides are bound at their sites of origin. Therefore the last column of Table X also lists the quantity of emanation to be expected per gram of rock or soil.

Table X

Mean Content of Long-Lived Parent Nuclides of Natural Radioactive Decay Series on the Continental Crust*

Nuclide	Concentration	
	in [g/g]	*in [Curie/g]*
U^{238}	$3 \cdot 10^{-6}$	$1.0 \cdot 10^{-12}$
U^{235}	$2 \cdot 10^{-8}$	$4.3 \cdot 10^{-14}$
Th^{232}	$1 \cdot 10^{-5}$	$1.1 \cdot 10^{-12}$

*Israel, 1962.

Each of the three natural radioactive decay series, for which the decay schemes are shown in Figure 50, contain a gaseous intermediate product of rare gas character in the corresponding emanation (Em). The radioactive cycle begins with the release of a fraction of the emanation in the uppermost rock and soil layers to the air present in rock cracks and soil capillaries. The emanation content of the soil air is several orders of magnitude higher than that of the atmosphere. Consequently, a strong concentration gradient exists

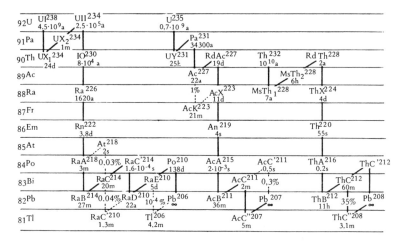

Figure 50. Decay scheme of the three natural radioactive decay series with isotope half lives. a = years, d = days, m = minutes, s = seconds.

between soil and air, leading to a diffusion current of the released emanations into the atmosphere. This release to the atmosphere, which is known as exhalation, leads to a depletion of emanations in the uppermost soil layer. Consequently, the emanation concentration increases with depth from a very small value at the surface of the earth until it reaches equilibrium with the parent nuclide.

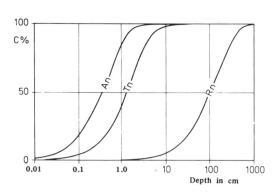

Figure 51. Depth variation of the three emanations in soils of constant activity and emanating power with an assumed diffusion coefficient $D = 0.05$ cm²/sec in the soil air.

Figure 51 shows the depth increase of the concentration c of the three emanations to be expected, expressed in per cent of the equilibrium value. Tn and An reach their end value already at a depth of a few centimeters, while Rn attains equilibrium only at a depth of a few meters because of its long mean life. The depth increase of the radon concentration has been confirmed by numerous measurements, while a corresponding confirmation for thoron and actinon has not been possible with the instrumentation available today.

It can now be assumed that on the average the atmospheric radioactivity content neither increases nor decreases. Consequently, the emanation content of the atmosphere must be equal to the depletion of the subjacent soil layer and the atoms disintegrating per unit of time must be resupplied by exhalation. These considerations lead to the following relation for the emanation concentration of an air column of unit cross section (see, for example, Israel, 1962):

$$G = a \sqrt{\frac{D}{\lambda^3}} \qquad (42)$$

where a is the emanation content corresponding to the radioactive equilibrium in soil air and D is the diffusion coefficient in the soil. If we use the equilibrium values from the third column of Table X for a, we obtain the ratio of the various emanation quantities in the atmosphere in curies:

$$G_{Rn} : G_{Tn} : G_{An} = 100 : 1.38 : 0.015 \qquad (43)$$

The decay products of ^{235}U (An-daughter products) thus are much less abundant than those of ^{238}U and ^{232}Th in the atmosphere and can therefore generally be neglected.

The emanation concentration in the soil air and the exhalation quantities resulting from them depend not only on the soil content of the parent nuclide but also on the emanating power of the soil material. Consequently, we find lower Rn-concentrations in the soil air than would be expected according to Table X. Table IX shows a summary of the mean Rn contents to be found on the continents. A comparison between these measurements with expected values indicates a mean emanating power of 10% for the soil material and a diffusion coefficient of about $D = 0.05$ cm^2/sec for the emanation. Exhalation measurements on continental soils have led to an average radon exhalation of $0.4 \cdot 10^{-16}$ C/cm^2 sec (Israel, 1962) and for thoron $40 \cdot 10^{-16}$ C/cm^2 sec (Grozier, 1969).

After exhalation from soil capillaries, the emanations reach the zone of atmospheric air motion. Thus, gaseous diffusion in the surface of the earth is replaced by the eddy diffusion in the atmosphere and by transport by air currents.

Since the soil represents a surface source of activation, for the moment we will neglect horizontal transport produced by wind in our orienting examination of the altitude-dependence of radioactivity over the continent, and will limit ourselves to vertical distribution due to mass transfer by eddy diffusion. The basis for our mathematical treatment of the altitude dependence of emanations and their daughter products is furnished by the diffusion equations that appear in the following form here. For any member of the decay series, we have

$$\frac{\partial c_i}{\partial t} = \frac{\partial}{\partial z}\left(K\frac{\partial c_i}{\partial z}\right) - \lambda_i c_i + \lambda_{i-1} c_{i-1} \tag{44}$$

c = Radioactivity concentration
z = Altitude
t = Time
λ_i = Decay constant
λ_{i-1} = Decay constant of parent nuclide
K = Eddy diffusion coefficient

with the boundary conditions for altitude 0:

$$c_1(0) \neq 0$$
$$c_i(0) = 0 \quad \text{for} \quad i > 1$$

Index 1 refers to the respective emanation, and indices 2, 3, ..., i, ... refer to the daughter products. Relation (44) is derived from the fact that the change of a radionuclide with time must correspond to the balance of resupply by forward diffusion and disintegrating parent nuclide, and loss by radioactive decay. The boundary conditions at the surface of the earth take into account that only emanations are exhaled and that the surface of the earth can with good approximation be considered a sink for the inductions.

If we neglect changes with time, *i.e.*, if $\frac{dc}{dt} i = 0$, and we integrate Equation (44), we obtain the stationary altitude distribution of individual nuclides. Figure 52 shows the result of such an integration for an apparent diffusion coefficient of plausible value which increases linearly from the surface up to a 1000 m altitude and then increases independent of altitude. The diagram shows

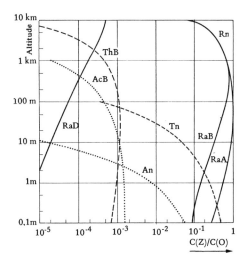

Figure 52. Altitude distribution of radioactive emanations and inductions under the conditions described in the text (according to Jacobi, 1960).

the ratio $c(z)/c(0)$ of the concentration at altitude z to the emanation at altitude 0 (Jacobi, 1960).

The emanations decrease continuously with altitude from their maximum value at the surface. The half value altitude for radon amounts to about 1000-2000 m, while the thoron content already decreases to 10% in the lowest meters of the atmosphere. The daughter products increase from a value of 0 at the surface to a maximum and then decrease again with increasing altitude. Provided they have a shorter half life than the emanations, radioactive equilibrium is reached at a certain altitude (Rn-RaA-RaB). In contrast, if they are longer-lived, a different altitude distribution of emanation and induction results (Tn-ThB). The specific relationships between natural radioactivity and atmospheric processes are discussed in Chapter 4, page 109).

Production of Radionuclides by Cosmic Rays. The next group covers the production of radionuclides by cosmic radiation. Cosmic radiation, which originates from space and continuously bombards the atmosphere, consists of 91.5% protons (hydrogen nuclei), 7.8% α-particles and 0.7% heavy nuclei with atomic numbers of up to about 30. On entry into the atmosphere, these extraordinarily high-energy primary particles collide with atmospheric gas particles and initiate nuclear reactions, thus releasing several nucleons (neutrons and protons), which in turn interact with additional nuclei. With further penetration into the atmosphere, a nucleon cascade is generated, and it is the protons and neutrons of this process that are mainly responsible for the production of new radionuclides in the atmosphere (and lithosphere). The energy of individual nucleons deep in the cascade decreases continuously until the particles reach a state of rest in the surrounding air and have only thermal energy due to Brownian motion.

Nitrogen nuclei have a high affinity for neutrons in the thermal energy interval. As a consequence, thermal neutrons of the nucleon cascade are captured almost exclusively by atmospheric nitrogen and thus lead to the formation of radioactive carbon ^{14}C (see footnote 1). The resulting mean overall ^{14}C production can be estimated at about 2.5 atoms per cm^2 of the earth's surface and per second. In the atmosphere, ^{14}C is oxidized relatively rapidly into ^{14}CO and then ^{14}CO$_2$, subsequently participating in the CO$_2$ cycle (see Chapter 2, page 8).

The formation of all other radionuclides can be attributed exclusively to the high-energy interaction of the nucleonic component with atomic nuclei of the atmospheric components. Production takes place by the so-called process of spallation, a nuclear process in which a very high energy injection excites the atomic nucleus to such a level that it ejects several nucleons simultaneously. Spallations can be detected, for example, on nuclear emulsions or in Wilson cloud chambers. The emitted nucleons leave star-shaped tracks originating from the excited nucleus. Therefore the spallation process is often called star production.

As a result of the terrestrial magnetic field, electrical interactions occur between the charged primary and secondary cosmic ray particles and the magnetic field. This leads to a geomagnetic latitude-dependence of the cosmic radiation intensity and thus of star production. Figure 53 shows the estimated dependence of star production on latitude and altitude according to Lal and

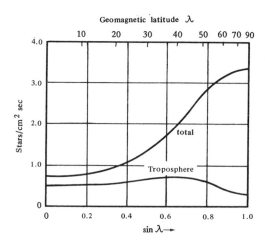

Figure 53. Number of stars per cm^2 of earth's surface and second as a function of the geomagnetic latitude in the years 1948/1949 (according to Lal and Peters, 1967).

Peters (1967). It indicates that in the troposphere star production varies little with geomagnetic latitude, amounting to about 0.5 per cm^2 and sec. In the stratosphere, where about 70% of the stars are produced, production increases with the geomagnetic latitude. At an altitude of 20-30 km star production reaches a maximum and then decreases with increasing altitude because of the decreasing density of the atmosphere (Lal and Peters, 1962).

The global mean of star production is estimated to be 1.8 spallations per square centimeter and second. Star production is reduced with increasing frequency of sun spots. Consequently, time variations of 10-20% are observed in the spallation rate during a sun spot cycle of 11 years. In accordance with the composition of the atmosphere, the stars are distributed among nitrogen, oxygen and argon in a ratio of about 76.5 : 22.5 : 1%. The radioactive spallation products, their mean global production rates and half lives are compiled in Table XI. At the lowest altitude up to 15 km the production of each of these elements is directly proportional to star production. At greater altitudes, however, this law shows certain deviations (see Lal and Peters, 1967).

Table XI

Production of Radionuclides by Cosmic Radiation in the Atmosphere*

Element	Half Life	Production Rate	
		$Atoms/cm^2 sec$	$Curie/cm^2 sec$
Hydrogen, ^3H	12.3 a	0.25	$1.3 \cdot 10^{-20}$
Carbon, ^{14}C	5760 a	2.5	$2.5 \cdot 10^{-22}$
Beryllium, ^7Be	53 d	$8.1 \cdot 10^{-2}$	$3.2 \cdot 10^{-19}$
Beryllium, ^{10}Be	$2.5 \cdot 10^6$ a	$4.5 \cdot 10^{-2}$	$9 \cdot 10^{-27}$
Sodium, ^{22}Na	2.6 a	$8.6 \cdot 10^{-5}$	$1.7 \cdot 10^{-23}$
Silicon, ^{32}Si	710 a	$1.6 \cdot 10^{-4}$	$1.3 \cdot 10^{-25}$
Phosphorus, ^{32}P	14 d	$8.1 \cdot 10^{-4}$	$1.2 \cdot 10^{-20}$
Phosphorus, ^{33}P	25 d	$6.8 \cdot 10^{-4}$	$6.8 \cdot 10^{-21}$
Sulfur, ^{35}S	87 d	$1.4 \cdot 10^{-3}$	$3.5 \cdot 10^{-21}$
Chlorine, ^{36}Cl	$3.5 \cdot 10^5$ a	$1.1 \cdot 10^{-3}$	$2.2 \cdot 10^{-27}$
Chlorine, ^{39}Cl	55 m	$2.3 \cdot 10^{-4}$	$1.3 \cdot 10^{-18}$

*According to Lal and Peters, 1967. a = years, d = days.

Artificial Radioactivity. Since the early fifties, artificial radio-activity, which is produced in uranium and plutonium fission processes and which can enter the air, water and soil by nuclear weapons tests, must be added to the preceding two groups of natural radionuclides. If certain elements with a large number of nucleons, such as uranium with a mass number of 235, are irradi-ated with neutrons, the nucleus is fissioned into two fragments during reaction with neutrons. In addition, some new neutrons are released. These two fragments have a different size and have a mass ratio of about 2 to 3 (see Figure 55, page 111). In addition, a large amount of energy is released during the fission process.

Neutrons released in the fission process initiate additional fis-sions provided a sufficient quantity of fissile material is available. This is the basis of the chain reaction for the generation of power as well as the chain reaction in the detonation of nuclear weapons.

Practical significance for the utilization of nuclear fission thus far has been attained only by the uranium isotopes ^{233}U and ^{235}U and by the plutonium isotope ^{239}Pu. The frequency distribution of the fission products of two of these isotopes is shown in Figure 54. The ordinate indicates the number of fragments of a given mass number (sum of pro-tons and neutrons in the nucleus) produced on the average in one hundred nuclear fission events.

Figure 54 shows that the fission yield has two maxima of about 6.5%, found at the mass number of 95 and 140 for ^{235}U. As a rule, fission prod-ucts are radioactive and after a few inter-mediate conversions they become stable elements. Table XII shows a part of the "artificial decay series" produced in this pro-cess.

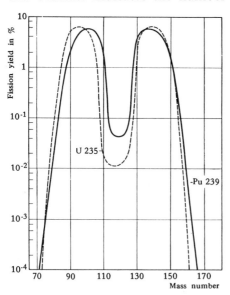

Figure 54. Fission yield in per cent of fissioned atoms. Solid curve: ^{239}Pu, dashed curve: ^{235}U (Riezler, Walcher, 1958).

Table XII

Summary of a Part of Artificial Radioactive Fission Products as a Decay Scheme*

Decay chains by mass number A (elements across: Tin, Antimony, Tellurium, Iodine, Xenon, Cesium, Barium, Lanthanum, Cerium, Praseodymium, Neodymium, Promethium, Samarium, Europium, Gadolinium, Terbium, Dysprosium; left diagonal labels: 50 Sn, 51 Sb, 52 Te, 53 J, 54 Xe, 55 Cs, 56 Ba, 57 La, 58 Ce, 59 Pr, 60 Nd, 61 Pm, 62 Sm, 63 Eu, 64 Gd, 65 Tb, 66 Dy):

A	Decay chain
131	3,4m → 23m → 25m → 8d → stab.
132	2,2m → 2,1m → 78h → 2,3h → stab.
133	44m → (2m) → 21h → 5,3d → stab.
134	50s → 44m → 53m → stab.
135	<2m → 6,5h → (15m) → 2,1 Ma → stab.
136	1,5m → stab. → 12,9a → stab.
137	22s → 3,4m → 30a → stab.
138	5,9s → 17m → 32m → stab.
139	2,7s → 41s → 9,5m → 8,5m → stab.
140	16s → 66s → 12,8a → 40h → stab.
141	1,7s → kurz → 18m → 3,7h → 33d → stab.
142	1m → 6m → 74m → stab.
143	1s → kurz → <0,5m → 19m → 33h → 13,9h → stab.
144	~1s → kurz → kurz → kurz → 290d → 17,5m → stab.
145	0,8s → kurz → kurz → kurz → 3m → 6h → stab.
146	14m → 25m → stab.
147	11d → 2,5a → stab.
148	stab.
149	2h → 50h → stab.
150	stab.
151	15m → 27,5h → 93a → stab.
152	stab.
153	<5m → 47h → stab.
154	stab.
155	<5m → 24m → 1,7a → stab.
156	<5m → ~10h → 14d → stab.
157	15,4h → stab.
158	1h → stab.
159	18h → stab.
160	stab.
161	7d → stab.

Yield in % of fissioned nuclei:

A	Slow neutrons U^{233}	U^{235}	Fast neutrons Pu^{239}	U^{238}
131	2,7	2,9	3,6	
132	4,9	4,3	4,9	
133		6,5	5,0	
134		7,5		
135		6,3	5,5	
136	1,7	6,2	1,9	
137		5,9		
138		5,7		
139		6,2	5,4	5,1
140	6	6,4	5,4	5
141		5,7	4,9	
142		5,9		
143		6,2		
144	3,4	6,0	3,7	
145		4,0		
146		3,2		
147	0,6	2,6		
148		1,8		
149		1,3		
150		0,7		
153	0,06	0,14	0,39	
155		0,031	0,21	
156		0,013		
157		$74 \cdot 10^{-3}$		
158		$2 \cdot 10^{-3}$	0,12	0,063
159		$1,1 \cdot 10^{-3}$		
161		$8 \cdot 10^{-5}$		

*Riezler, Walcher, 1958. A = mass number.

Most conversions take place by β-decay becoming evident by the absence of a change in mass number A during disintegration, while the number of nuclear protons increases by one. The number of nuclear protons is indicated on the inclined lines in Table XII for the corresponding elements.

The radioactive data of some important fission products that either are produced in particular abundance or are of special significance because of their long half life are listed in Table XIII. If such a fission product mixture is introduced into the atmosphere, for example, by a nuclear weapon explosion, its total activity decays in the first three months approximately according to the relation (Hunter, Ballou, 1951)

$$A = A_0 t^{-1,2} \tag{45}$$

A_0 = fission product activity at
the time of detonation, *i.e.*, t = 0

Compared to the decay law of a single substance (Equation 39b) this behavior differs because here we are dealing with a mixture of elements with the most diverse decay constants. In the course of time, the composition of the fission product mixture shifts increasingly in favor of the longer-lived nuclides. Table XIV shows the anticipated percentage composition as a function of the age of the mixture (maxima in italics). We can recognize the change from the preponderant content of ^{140}Ba and ^{140}La during the first 40 days to ^{95}Zr and ^{95}Nb up to 250 days, then to ^{144}Ce and ^{144}Pr, and finally to ^{90}Sr and ^{137}Cs after a number of years.

The quantity of fission products formed in an explosion of a weapon based on nuclear fission (A-bomb) is directly proportional to the explosive force of the weapon, which usually is expressed in equivalents of ordinary TNT (kt = 1000 tons, Mt = million tons). In the detonation of a 1 kt weapon, approximately $1.45 \cdot 10^{22}$ atoms are split. In a fusion weapon (H-bomb), large quantities of tritium are added to the products originating from the fuse.

In addition to the actual fission products, activities are induced in materials of the surrounding region by the released neutrons, such as in the construction material of the weapon, in entrained soil material, or in other materials in the range of the detonation.

A number of these induced activities are nuclides that are already present in nature due to production by cosmic radiation. Thus, natural ^{36}Cl can hardly be detected anywhere any longer

Table XIII

Radioactive Properties of Some Important Fission Products*

A	Element	$t_{1/2}$	E_β	E_γ	%
85	Kr	4.4 h	0.85	0.15	1.2
85	Kr	10.6 a	0.7	—	0.32
87	Kr	1.3 h	3.2	—	2.7
88	Kr	2.77 h	2.7 u. a.	2 & others	3.6
	Rb	17.7 m	5.3 (k)	1.85 & others	
89	Sr	51 d	1.5	—	4.8
90	Sr	28 a	0.6	—	5.8
	Y	65 h	2.2	—	
91	Sr	9.7 h	3.2 (k)	1.41 & others	
	Y	50 m	—	0.551	
	Y	57 d	1.5	—	5.8
92	Sr	2.7 h	—	—	5.0
	Y	3.4 h	3.6	0.94 & others	6.0
93	Y	10 h	3.1	0.7	6.3
	Zr	0.95 Ma	0.063	—	
95	Zr-	65 d	0.37	0.73	6.4
	Nb	35 d	0.16	0.95	
96	Nb	23 h	0.750	0.77 & others	6.3
97	Nb	1.2 h	1.27	0.66	6.3
99	Mo	68 h	1.23	0.04 - 0.78	
	Tc	6 h	—	0.14	6.1
	Tc	0.22 Ma	0.29	—	
103	Ru	40 d	0.22	0.50	2.9
	Rh	57 m	—	0.04	
106	Ru	1 a	0.039	—	0.38
	Rh	30 s	3.5 (k)	0.5 & others	
107	Pd	7 Ma	0.04	—	(0.1)
129	Te	1.2 h	1.8	0.3	1.0
	J	30 Ma	0.15	0.038	
131	J	8 d	0.6 (k)	0.36 & others	2.9
132	Te	78 h	0.22	0.23	4.3
	J	2.3 h	—	—	
133	Xe	5.3 d	0.34	0.08	6.5
135	J	6.65 h	1.4 (k)	1.38 & others	
	Xe	9.2 h	0.91	0.25	6.3
	Cs	2.1 Ma	0.21	—	
137	Cs	30 a	0.52	—	5.9
	Ba	2.6 m	—	0.66	
139	Ba	85 m	2.38 (k)	0.165	6.2
140	Ba	12.8 d	1.02 (k)	0.16 - 0.54	6.4
	La	40 h	2.2	0.09 - 2.5	
141	La	3.7 h	2.43	—	5.7
	Ce	33 d	0.58	0.14	
142	La	74 m	2.5	0.63; 0.87	5.9
143	Ce	33 h	1.40 (k)	0.057 - 1.10	
	Pr	13.9 h	0.93	—	6.2
144	Ce	290 d	0.30 (k)	0.3 - 0.13	6.0
	Pr	17.5 m	2.97	2.2 & others	
145	Pr	6 h	1.7	—	4.0
147	Nd	11 d	0.8 (k)	0.52 & others	2.6
	Pm	2.5 a	0.22	—	
149	Nd	2 h	1.5 (k)	0.030 - 0.65	1.3
	Pm	50 h	0.97	0.285	
151	Sm	93 a	0.076	0.019	(0.5)
155	Eu	1.7 a	0.24 (k)	0.018 - 0.102	0.031

*A = mass number, $t_{1/2}$ = half life (a = years, Ma = million years, d = days, h = hours, m = minutes). E_β = maximum energy of the β-spectrum; E_γ = energy of γ-quanta in MeV. Per cent: % frequency of fissioned ^{235}U atoms (Riezler, Walcher 1958).

Table XIV

Contribution in Per Cent to the Fission Product Activity of a Mixture Formed by ^{235}U Fission as a Function of Age (preponderant contributions emphasized in italics)*

Element	Age in Days												
	10	20	30	40	50	60	80	100	150	200	250	300	365
Ba^{140}, La^{140}	*22.6*	*25.9*	*23.2*	*18.9*	14.6	11.0	5.4	2.2					
Xe^{133}	11.5	6.0	2.5	1.0									
Te^{132}, I^{132}	10.4	2.6											
Pr^{143}	10.0	12.0	11.2	9.6	7.8	6.2	3.4	1.4					
I^{131}	6.8	5.6	3.6	2.3	1.4								
Mo^{99}	6.8	1.1											
Ce^{141}	6.3	9.7	11.2	11.7	11.3	10.8	9.3	7.8	4.5	1.8			
Ru^{103}, Rh^{103}	5.1	8.6	11.2	13.0	14.0	14.6	14.6	13.8	10.5	7.1	4.5	2.9	
Nd^{147}, Pm^{147}	4.8	5.0	4.1	3.1	2.1	1.4			1.4	2.2	3.2	4.2	5.8
Zr^{95}, Nb^{95}	3.3	8.1	12.2	16.6	*20.5*	*24.5*	*30.3*	*34.5*	*39.2*	*39.4*	*35.4*	*29.5*	*22.5*
Y^{91}	3.2	5.6	7.6	9.0	10.2	11.2	12.3	12.8	12.0	10.3	9.5	6.5	3.8
Sr^{89}	2.9	5.0	6.7	8.0	9.0	9.8	10.3	10.5	9.6	7.9	6.1	4.5	2.7
Ce^{144}, Pr^{144}		2.6	4.0	5.4	6.6	8.0	10.6	13.4	19.6	26.0	33.6	*42.0*	*53.0*
Ru^{106}, Rh^{106}											3.0	3.8	4.8
Sr^{90}, Y^{90}											2.0	2.6	3.8
Cs^{137}, Ba^{137}												2.0	3.0

Element	Age in Days										
	1	2	3	4	6	8	10	15	20	50	100
Zr^{15}, Nb^{95}	22.0	1.0									
Y^{91}	3.8										
Sr^{89}	2.7										
Ce^{144}, Pr^{144}	*53.9*	*60.0*	*41.0*	*27.0*	8.6	1.9					
Pm^{147}	5.8	13.5	19.0	21.3	21.2	18.8	15.5	8.0	3.4		
Ru^{106}, Rh^{106}	4.8	6.8	6.0	4.7	2.3						
Sr^{90}, Y^{90}	3.8	10.4	17.4	24.0	*34.0*	*40.0*	*43.8*	*47.6*	*48.5*	*44.0*	*34.0*
Cs^{137}, Ba^{137}	3.0	8.0	13.6	18.8	27.4	33.0	36.6	41.8	44.0	*54.5*	*64.0*
Sm^{151}			1.2	1.5	2.0	2.3	2.5	2.6	2.6	2.0	1.1
Kr^{85}			1.2	1.5	1.7	1.7	1.5	1.2			

*According to Hunter, Ballou, 1951.

compared to that produced artificially above the ocean during explosions. As a result of neutron capture by atmospheric nitrogen, large quantities of ^{14}C are also formed. Measurements also indicate that the isotopes ^{35}S, ^{22}Na and ^{7}Be are produced occasionally.

The induced activities also include elements that do not occur or have not yet been found in nature, *i.e.,* the transuranium compounds. These elements, whose atomic number is higher than 92 (uranium), are produced from the weapon material by nuclear reactions. In the case of fusion weapons explosions, elements with atomic numbers of up to 100 (fermium) have been detected.

Activities that are frequently formed in the construction material of the weapon include mainly cobalt, iron, manganese, zinc and tin isotopes. In some cases, these are joined by materials that were produced intentionally by additives to the weapons material in order to allow observation of the products formed in a given explosion. They particularly include rhodium-102, cadmium-109 and tungsten-185.

The Atmospheric Radioactivity Cycle

In the discussion of atmospheric radioactivity thus far we stressed its production and decay characteristics, while questions concerning its dispersion and distribution in the atmosphere and its significance as an atmospheric component were only briefly mentioned. Therefore these questions will be considered in detail now.

The dispersion of gaseous radioactive materials or those added to aerosols in the atmosphere is governed by the same physical and chemical laws responsible for the dispersion of trace gases and aerosols. An important difference compared to inactive materials, however, consists of radioactive decay, which represents an additional removal mechanism and thus modifies the history of these elements in the atmosphere.

In recent decades, radioactivity has become of particular importance for the investigation of atmospheric transport processes. Atmospheric circulation can be observed by means of tracer substances, provided their sources and sinks are known. Radioactive tracers have the advantage over inactive substances in that their sinks often consist exclusively of radioactive decay and thus are known. The availability of numerous materials with different half lives and origins allows their use for studies of the entire atmospheric transport process from microturbulence to global exchange.

Radon, Thoron and Daughter Products. The atmospheric cycle begins with the emission of Rn and Tn from the surface of the earth. Exhalation is controlled by various factors in addition to the uranium and thorium content of the soil, and these factors can cause variations of about 200-500%. In particular, there is an increase of exhalation with wind speed, probably attributable to the vacuum effect of wind on soil capillaries. Precipitation and increasing soil moisture lead to a reduction of exhalation, presumably due to plugging of some of the soil capillaries. Other factors, such as pressure variations and temperature gradients probably play a considerably less important role (compare, for example, J. E. Pearson, 1965, M. Horbert, 1969, D. Guedalia *et al.,* 1970, Grozer, 1969).

As mentioned above, the surface of the continents are sources of radon and thoron. Whether the surface of the ocean and polar ice regions release emanations to the atmosphere or whether the Rn concentrations found there (Table IX) are of continential origin has not yet been determined (see H. Israel, 1951, L. B. Lockhart, Jr., R. L. Patterson, Jr. and A. W. Sanders, 1966). Thus, it may be expected that the longer-lived isotopes will show a dependence on wind direction. This is the case as shown in Figure 55.

This figure shows the "activity wind roses" for radon and its short-lived decay products, which are in radioactive equilibrium for different locations in Germany and Italy. The difference is most apparent between the continental and sea wind in Naples. Furthermore, the wind roses in northern Germany show the lowest values when winds come from the sea. With increasing distance from the coast, for example, in Berlin and Dresden, these differences become smaller, since the maritime air masses absorb radon as they traverse the continent and with time reach the continental saturation value. In the case of the wind roses found in southern Germany, regional exhalation differences might play an additional role which may possibly be attributed to the uranium regions in the southeast of central Europe—Erzgebirge, Sudetenland.

Because its half life is only 54 sec, atmospheric thoron, as may be expected, reacts only to influences in the closer vicinity, which represents a circumference of approximately a few hundred meters. Thus, measurements on homogeneous terrain with a uniform exhalation indicate as a rule no dependence on wind direction. In contrast, the thoron content may vary with changes in the height of vegetation such as grain, for example, during excavation work (H. Israel, Horbert, De la Riva, 1967). Wind roses for the ThB content have an appearance similar to the radon wind roses with a smaller amplitude of variation as is to be expected in accordance with the shorter half life of ThB, *i.e.,* 10.6 h compared to 3.8 days (see, for example, H. Israel and S. Stiller, 1963).

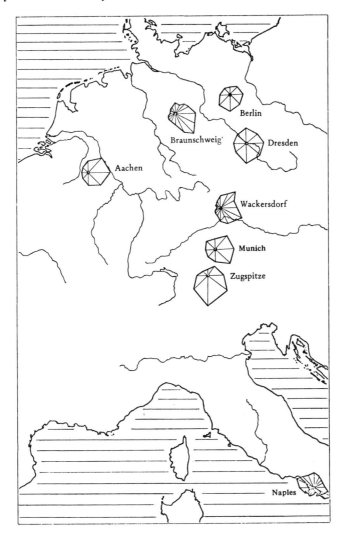

Figure 55. Dependence of radon and its short-lived decay products on wind direction at different locations in Germany and Italy (Israel, 1966).

The examined nuclides are subject to periodic as well as aperiodic variations under meteorological influences. Generally, the atmospheric Rn and ThB concentration exhibits diurnal variations with maxima during the night and near the morning and minima in the afternoon hours. The diurnal fluctuations are more pronounced on days with great variations in vertical exchange than on days when these are minor, which suggests a close relationship between diurnal fluctuations and the exchange process. In addition to exchange, the above-mentioned local exhalation fluctuations naturally also play a decisive role. In mountain areas, these variations are less pronounced and evidently are decisively influenced by the exchange process. Thus, clearly defined regular diurnal fluctuations at mountain stations usually occur only in the presence of a strong convective air exchange with lower layers. The annual fluctuation of the atmospheric Rn content is nonuniform because several controlling factors overlap in this process.

The stationary altitude distribution of the emanations is determined by vertical exchange and radioactive decay provided exhalation is uniform (compare Equation (44) and Figure 52). Consequently, a half-value altitude of only a few meters is expected for thoron, while the half-value altitude of Rn should be at about 1000 m. This has been confirmed experimentally as shown by the examples in Figure 56.

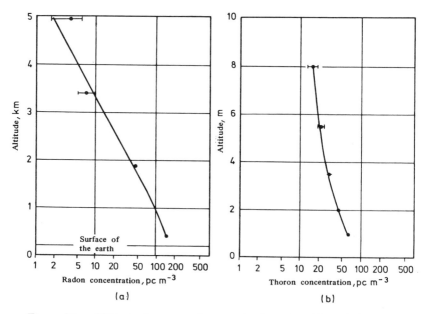

Figure 56. Altitude decrease of atmospheric radon (a) and thoron concentrations (b). |—•—| = measured altitude profiles; solid curve = calculated according to Equation (44) with the assumption of the trend of the apparent diffusion coefficient K as described in the text.

The test points in Figure 56 measured over Illinois show a typical altitude profile for radon (Bradley, Pearson, 1970). The solid curve has been calculated for the stationary case according to Equation (44) (Jacobi, Andre, 1963) with the assumption of normal exchange at ground level and a rapid altitude decrease of the apparent diffusion coefficient between 100 and 1000 m. Figure 56b shows a similar comparison for a mean thoron profile in Mainflingen (Israel, 1965). It can be seen, that with the assumption of a suitable altitude trend of the apparent diffusion coefficient, agreement between theory and experiment can be obtained in these examples. Thus, this method makes it possible to use radioactivity measurements to determine the eddy diffusion coefficient, which characterizes atmospheric vertical exchange. It is true that the applicability of the method is limited to cases in which the assumptions serving as the basis of the theory (stationary conditions with uniform exhalation) are satisfied in nature. Admittedly this is often not the case. An example of nonuniform exhalation in which horizontal air mass transport plays a decisive role is given by the radon measurements of Birot *et al.* (1965) over southern France. They found that after a rapid initial altitude decrease, the radon concentration increased again showing a maximum at about 2 km altitude. The authors were able to attribute this behavior to air masses of different origin. The air near ground level was of marine origin with a relatively low radon content, while the upper air layers were of continental origin and therefore had a higher initial radon concentration.

The short-lived daughter products of radon (RaA, RaB, RaC and RaC') and thoron (ThA) are practically in radioactive equilibrium with the corresponding emanation because of their short half lives and therefore show essentially the same altitude distribution.

The situation is different for the longer-lived daughter products, among which RaD, RaE, RaF and ThB are of particular interest for exchange studies in the troposphere and stratosphere. These heavy metals collect on the atmospheric aerosol, so that their behavior comes to be closely related with the aerosol cycle. This relationship is shown schematically in Figure 57.

In addition to radioactive decay, therefore, a further decrease of the inductions occurs as a result of the removal of carrier aerosols by sedimentation and precipitation. A part of the as yet undecayed inductions is thus returned to the surface of the earth. Consequently, the radium-D-, E- and F-content of the atmosphere

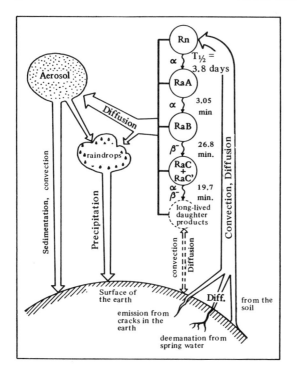

Figure 57. Diagram of the relationship of atmospheric radioactivity with the aerosol cycle (Jacobi, 1960).

is much lower than would be expected on the basis of the radioactive equilibrium. This decrease relative to the equilibrium value or the ratio of short- to long-lived activities can furnish information on the mean aerosol residence time.

Discussions and measurements of this type in regard to rain and air samples have been published by several authors (see, for example, Junge, 1962) and will therefore only be briefly considered here. In order to obtain a qualitative idea, we will assume for the sake of simplicity that the atmospheric concentration of emanations and their daughter products is independent of altitude. Furthermore, we assume that aerosol deposition can be described by the same laws as radioactive decay (Equation 39a). The decay law for the atmospheric decay series is then obtained by a suitable expansion of Equation (40):

$$\frac{dc_i}{dt} = \lambda_{i-1} c_{i-1} - \lambda_i c_i - \lambda_a c_i; \quad i = 1, 2, 3 \ldots \quad (46)$$

where λ_a is the decay constant of the aerosol. We let $\lambda_a = 0$ for the emanation $i = 1$, since the emanations do not participate in the aerosol cycle.

If the radon concentration c_{Rn} is assumed to be constant in time, this results in the following for the equilibrium:

$$c_i = c_{Rn} \lambda_{Rn} \cdot \frac{\lambda_1}{\lambda_1 + \lambda_a} \cdot \frac{\lambda_2}{\lambda_2 + \lambda_a} \cdots \cdot \frac{\lambda_{i-1}}{\lambda_{i-1} + \lambda_a} \cdot \frac{1}{\lambda_i + \lambda_a} \quad (47)$$

For short-lived daughter products RaA to RaC, we have $\lambda_i \gg \lambda_a$ and the relation simplifies into the well-known expression for radioactive equilibrium:

$$\lambda_i c_i = \lambda_{Rn} c_{Rn} \quad (41)$$

In contrast, for long-lived nuclides, such as RaD, which has a very small decay constant compared to λ_a, we obtain the relation:

$$c_{RaD} = \frac{\lambda_{Rn} c_{Rn}}{\lambda_a} \quad (48)$$

and

$$\lambda_a = \frac{\lambda_{Rn} c_{Rn}}{c_{RaD}} = \frac{\lambda_{RaB} c_{RaB}}{c_{RaD}}$$

Expression (48) can be used to estimate the mean aerosol lifetime $\tau_a = 1/\lambda_a$ by determining c_{RaD} and c_{RaB} by means of filter or rainwater samples and forming the ratio according to (48). These and similar measurements of other natural and artificial isotope ratios lead to tropospheric aerosol residence times of a few days in the lower layers and up to about one month in the upper troposphere.

Measurements of the tropospheric vertical profile of long-lived inductions offer a further opportunity to estimate the aerosol residence times. If Equation (44) is expanded by the member that accounts for the deposition of the carrier aerosol, we obtain the following in the case of equilibrium:

$$\frac{d}{dz}\ K\frac{dc_i}{dz}\ + \lambda_{i-1}\,c_{i-1} - (\lambda_i + \lambda_a)\,c_i = 0,\ i = 1, 2, 3 \dots \quad (49)$$

with $\lambda_a = 0$ for $i = 1$.

If this differential equation is solved with the assumption of different eddy diffusion coefficients K and aerosol residence times $1/\lambda_a$, then a family of theoretical vertical profiles is obtained for the activity, which can be compared with the measured profiles.

This method is illustrated in Figure 58 in which a mean RaD profile measured by Burton and Stewart (1960) over England is compared with calculated profiles. The observed concentration increase in the troposphere is in fairly good agreement with the calculated profiles for aerosol residence times of 30-70 days. The concentration increase above the tropopause suggests a much longer stratospheric aerosol residence time.

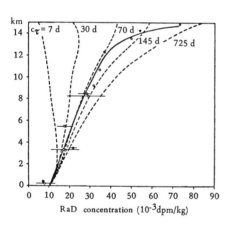

Figure 58. Solid curve and test points: mean RaD profile observed over England (Burton and Stewart, 1960); dashed curves: calculated RaD profile with the assumption of $K = 2 \cdot 10^5$ cm^2/sec in the troposphere and $K = 2 \cdot 10^3$ cm^2/sec in the stratosphere and different aerosol residence times τ_a (Jacobi, Andre, 1963).

Numerous measurements exist for RaD and other long-lived daughter products, and they have produced the same qualitative pattern. However, it should be pointed out once more that the methods for the aerosol residence time determination described here can furnish only approximate values, since the necessary assumptions to allow a mathematical treatment frequently oversimplify physical conditions and therefore are only approximations of reality.

For the stratospheric RaD distribution we obtain the following approximate picture (Figure 59). Immediately above the tropopause, a layer of several kilometers depth with a relatively high

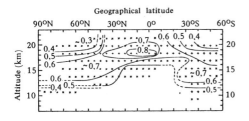

Figure 59. RaD distribution in the lower stratosphere in 10^{-14} curies per standard cubic meter (Freely *et al.*, 1970).

RaD content that follows the inclination of the tropopause from the equator to the pole is observed. The highest RaD values are found in the tropical stratosphere. Feely and Seitz (1970) conclude that the stratospheric RaD distribution may possibly be interpreted as an equilibrium state between stratospheric horizontal and vertical transport and sedimentation of the carrier aerosols; they conclude that RaD should be suitable as a tracer for a study of stratospheric exchange as well as of exchange between the troposphere and stratosphere.

Radioactivity Originating from Cosmic Radiation. Tritium and radioactive carbon are incorporated in the water and carbon dioxide cycle, respectively, and can be used for their study. The ^{14}C cycle has already been discussed in detail in Chapter 2, page 8 in connection with CO_2 transport. Tritium was produced in large quantities by fusion weapons in the fifties and sixties, so that atmospheric tritium at the present time is mainly of anthropogenic origin. Since practically no ^{3}H measurements are available from the period previous to the nuclear weapons tests, tritium will be discussed in the next section together with artificial radioactivity.

The half lives of ^{7}Be ($t_{1/2}$ = 53d), ^{22}Na (2.6 a), ^{32}P (14d), ^{33}P (25d) and ^{35}S (87d) are in the same order of magnitude as the process of large-scale atmospheric circulations with time. Therefore these isotopes are mainly suitable for meteorological studies. The shorter-lived nuclides seem appropriate for a study of tropospheric circulations, cloud formations and aerosol cycles, while the longer-lived ones are of particular interest for studies of slower mixing processes in the stratosphere (compare, for example, Lal and Peters, 1967, Bolin, 1962, Bhandari, Lal and Rama, 1966). The use of ^{35}S and ^{22}Na for such studies is made difficult because these substances are also produced artificially in nuclear weapons tests, so that a clear assignment of sources is often no longer possible.

We will turn first to the stratospheric distribution of radioisotopes. As mentioned, the stratospheric production of these elements depends on altitude as well as on geomagnetic latitude (see Figures 53 and 60). If no air motion or turbulent mixing processes would take place in the stratosphere, radioactive equilibrium between resupply and decay would develop everywhere and the distribution of the various nuclides in the stratosphere would correspond to star production. However, in the presence of air motion, these concentration ratios at a certain location may change by the influx or efflux of a part of the radionuclides. The deviations from equilibrium as well as the activity ratio of nuclides of different half lives allow us to draw conclusions concerning the prevailing stratospheric circulations (see, for example, Bolin, 1962, Lal and Peters, 1967).

In the last twenty years, the development of special filter materials has made it possible to obtain samples of atmospheric microaerosols from large air masses (up to several tons). This allowed a measurement of the atmospheric concentration of radionuclides produced by cosmic radiation. Sampling at altitudes of up to about 20 km is possible from planes. Since balloons or rockets must be used for greater altitudes, the available test material for altitudes above 20 km is still very limited.

Figure 60 shows an example in the form of a compilation of tropospheric and stratospheric ^{7}Be measurements made by Bhandari, Lal and Rama (1966). For a comparison, the areas of identical ^{7}Be production by cosmic radiation have been entered in the figure. Since beryllium production is proportional to star production, these areas at the same time represent a criterion of the altitude and latitude dependence of star production. The dashed line signifies the location of the tropopause, the temperature inversion, which forms the upper boundary of the troposphere to the stratosphere. The tropopause is primarily responsible for the very slow mass exchange between stratosphere and troposphere.

The measurements show that the calculated stratospheric ^{7}Be distribution is very similar to the measured data and that essentially equilibrium prevails between production and decay. The other short-lived isotopes, ^{32}P, ^{33}P and ^{35}S as well as the ratios ^{33}P/^{32}P, ^{7}Be/^{32}P and ^{7}Be/^{35}S calculated from these show a similar behavior and confirm that the stratospheric air mass displacements are so slow that these nuclides are practically always in equilibrium with production (Lal and Peters, 1967).

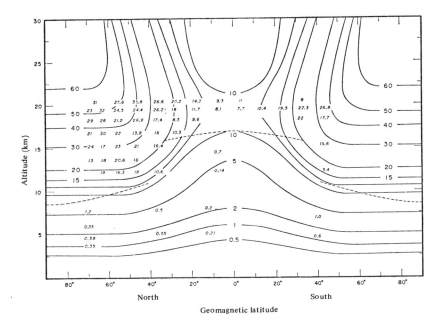

Figure 60. Profile of the mean atmospheric ⁷Be-concentration (disintegrations per minute and standard cubic meter). — 60 — = areas of equal ⁷Be production with indication of the expected equilibrium concentration (disintegrations per minute and standard cubic meter); ---- = position of tropopause (according to Bhandari, Lal and Rama, 1966).

²²Na measurements from the period prior to nuclear weapons show, in contrast, that the ²²Na concentration as well as the ²²Na/⁷Be ratio are considerably lower than the expected equilibrium value and that these deviations from equilibrium are highly dependent on altitude. With the assumption that vertical transport in the stratosphere is mainly the result of eddy diffusion, Bhandari *et al.* (1963 and 1966) calculated mean vertical eddy diffusion coefficients for different stratospheric regions and found values between 2 and 20 · 10^3 cm² /sec in agreement with fission product and ozone measurements.

Some studies indicate that significant seasonal fluctuations of cosmic radiation-produced nuclides occur in the stratosphere, which can possibly be utilized for a study of the annual fluctuations of stratospheric air motion. The recorded data, however, are too limited thus far to allow a definitive conclusion on the subject.

The tropospheric conditions differ considerably from the stratospheric. The rapid and variable tropospheric air motions prevent the development of stationary radionuclide distributions (see Figure 60). Since these nuclides are attached to the atmospheric aerosol, their behavior is also closely related with the tropospheric aerosol cycle. They are consequently removed from the atmosphere by precipitation with rainout and washout.

Numerous measurements of rainwater activity are available for various geographical locations. The precipitation activity is subject to great local fluctuations. However, it has been found that the mean ^7Be, ^{32}P and ^{33}P concentrations do not exhibit a substantial dependence on geographical location. This is to be expected if the nuclides deposited with the precipitation are of tropospheric origin, since tropospheric star production shows very little latitude dependence.

Measurements of the concentration ratio of different isotopes in rainwater can offer information on the aerosol residence time relative to deposition by precipitation. We shall assume that the radioactivity of an air mass has been completely washed out by a rain. If we now follow this air mass, the various nuclides are slowly reformed by cosmic radiation until they are again washed out by the next precipitation. Since the irradiation time necessary to attain radioactive equilibrium increases with the half life of the various nuclides, the concentration ratio of the individual nuclides in the rainwater is a criterion for the exposure age of the air mass or a measure of the residence time of the aerosol in such an air mass. Measurements of the $^7Be/^{32}P$ ratio in rainwater show mean residence times or washout periods of about 40 days, in agreement with measurements of the aerosol residence times made on the basis of radon and thoron daughter products and nuclear fission products. The mean precipitation activity is also in agreement with the tropospheric production for a washout interval of 30-40 days.

Artificial Radioactivity. The sources of artificial radioactivity are the sites of nuclear fission and fusion weapons explosions. An explosion essentially forms radioactive aerosols in addition to a few gaseous radionuclides, for example, tritium and xenon. These particles of artificial radioactivity are produced by a basically different mechanism than natural radioactive aerosols.

If a nuclear bomb is detonated at sufficient altitude above ground level, the fireball contains mainly the diluted gases of iron and other construction materials, while the quantitative contribution of radioactive fission products is small. During cooling of the fireball, iron oxide particles of up to about 1 micron size form first of all by condensation and coagulation. Most of the nongaseous activities condense later when the fireball has cooled further and therefore are primarily deposited on iron oxide particles. In the case of an air blast, therefore, the aerosols produced are for the most part so small that they do not undergo significant sedimentation, consequently remaining suspended in the atmosphere until they are deposited by precipitation.

These relatively simple conditions of production are modified during surface blasts because considerable quantities of soil material are entrained into the fireball. As a result, particles of up to a few tenths of a millimeter size are present in the fireball, depending on the size of the weapon; a part of the activities is bound to these particles. Since aerosols with diameters of more than about 10 microns are deposited relatively rapidly by sedimentation, these explosions are accompanied by a particle rain of high activity close to the blast site, which is known as near fallout. A second group of particles with radii between 0.5 and 10 micron are transported over far regions of the earth by air currents. Their fallout is determined not only by sedimentation but to a considerable extent by turbulence, convection and precipitation. The aerosols of this so-called intermediate fallout, which may remain suspended in the atmosphere for up to a few weeks under certain conditions, can be recognized by their high specific activity which has led to their name of "hot particles." Particles with activities of up to 10^9 curie have been observed. In aerosols with a radius of less than 0.5 micron, deposition by sedimentation is negligible. They follow the atmospheric air motion until they are deposited on the surface of the earth by rainout or washout. The activity transported to the surface by them is called long-term delayed fallout.

In the case of nuclear blasts at the surface or at low altitude as well as tropospheric air blasts of not too high energy, the tropospheric component of the weapons products can be identified in the form of radioactive clouds. The movement of such clouds around the earth and their repeated detection at individual stations demonstrates that such a cloud can remain intact for several weeks. However, as a result of volume expansion, radioactive decay, and deposition processes, the activity content decreases with time.

In the case of explosions with power of more than 1 Mt and stratospheric detonations, almost the entire activity is transported into the stratosphere. Because of the stable stratification of the stratosphere and the consequent relatively slow vertical transport, the stratosphere represents a storage reservoir for long-lived fission products. The release of long-lived isotopes to the subjacent troposphere takes place slowly and shows an annual fluctuation with a spring maximum as well as a latitude dependence with a maximum in the middle latitudes (Lockhart *et al.,* 1960).

This is the result of the structure of the tropopause, the very stable air layer which forms the upper boundary of the troposphere toward the stratosphere. The tropopause has a complex structure and may show gaps, particularly in middle latitudes which allow an accelerated air mass exchange between these two atmospheric reservoirs. Figure 61 shows an example in the form of the annual fluctuation of strontium-90 in Heidelberg with a pronounced maximum in the spring. As expected, the ^{90}Sr content shows the highest values during or shortly after the extended weapons test series at the end of the fifties and in the early sixties, and it then slowly decays.

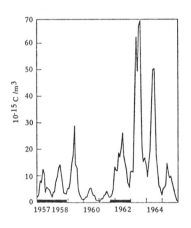

Figure 61. ^{90}Sr activity of the air near ground level in Heidelberg. The periods of nuclear weapons test activity are marked on the abscissa (according to Schumann, 1967).

The Americans have conducted some specific tracer experiments in order to investigate transport from high atmospheric layers to the surface of the earth. On 8-11-1958, three Megacuries of rhodium-102 ($t_{1/2}$ = 210 d) were injected into the atmosphere by a nuclear weapons explosion at 43 km altitude over Johnston Island in the Pacific (16°N; 170°W). During the following years, ^{102}Rh measurements were taken at 19.4 and 5.5 km altitude (Kalkstein, 1963). About 10 months after the explosion, ^{102}Rh increased at 20 km altitude over the Southern Hemisphere. A second rise occurred about one year later. In the Northern Hemisphere, a similar increase was observed after about 14 months. The slope and degree of increase both showed a pronounced latitude dependence and gave information on the horizontal transport in different layers of the stratosphere. An additional year passed until the tropospheric ^{102}Rh concentration reached its maximum. These observations lead to the conclusion that the atmospheric residence time of tracer materials in the upper atmosphere is in the order of three years. The results of a ^{109}Cd injection at about 400 km altitude led to similar findings. In addition to these specific experiments, world-wide measurements of fission products from nuclear weapons explosions have also contributed significantly to an understanding of large-scale global circulation.

More than 99% of the natural and artificial tritium is present in the form of tritiated water, HTO. It represents a tracer that is particularly suited for a study of the water cycle. The residual tritium is mainly bound in the form of HT and CH_3T.

5

Anthropogenic Sources of Atmospheric Trace Elements (Air Pollution)

In the discussion of trace gases and aerosols in the preceding chapters, repeated reference was made to the contributions of anthropogenic sources and the resulting modification of the atmospheric trace element content. Industry, motor vehicle traffic and the use of fossil fuels for home heating and energy generation furnish the main contribution to these artificial trace elements. The estimated annual emissions from various sources in the USA are listed in Table XV. Because of the extremely high motor vehicle density in the US, about 42% of the total emissions of 214 million tons originate from traffic. Coal and oil burning are in second place with 22% and industry in third place with 14%. The balance is attributable to various activities, for example, refuse incineration, open fires in agriculture, construction work.

The relative contributions presumably vary somewhat from one country to another depending on standard of living and geographical location. However, it can be considered a rule of thumb that

Table XV

Estimated Emissions of Anthropogenic Sources in the USA for 1968*

Source	CO	SO_x	NO_x	HC	Partic- ulates	Total Quantity
Traffic	63.8	0.8	8.1	16.6	1.2	90.5
Fuel consumption	1.9	24.4	10.0	0.7	8.9	45.9
Industry	9.7	7.3	0.2	4.6	7.5	29.3
Miscellaneous	24.7	0.7	2.3	10.7	10.1	48.5

HC = hydrocarbons; x = 1 or 2; SO_x and NO_x are expressed in SO_2 and NO_2.

*US Dept. HEW, 1968. Data in millions of tons.

each of the three main categories mentioned contribute approximately to an equal degree.

The artificially produced trace elements, such as the products listed in Table XV, provided they attain a sufficiently high concentration in the atmosphere near ground level, frequently are injurious to health and welfare, and lead to damage of vegetation and property. Thus, they represent noxious or toxic agents, which have become known by the general term of air pollutants.

As a result of continuously increasing demands for energy, advancing industrialization and motorization, air pollution today has reached intolerable levels in some areas of high population density in industrialized countries. Thus, in the US alone, the annual damage caused by air pollution is estimated to be $7-10 billion, although injury to public health and welfare, which cannot be expressed in terms of money, is not even taken into consideration. In isolated cases, pollutant concentrations have even attained levels that led to acute illness and death. Probably the most unfortunate events of this type occurred in London during particularly extended inversion conditions with fog. In November 1952, the number of deaths due to acute respiratory disease in a period of 14 days was about 4000 greater than in a comparable period of other years. This event recurred in 1956 with approximately 1000 excess deaths. However, elsewhere, for example, the Maastal, in Osaka, Japan, Donora, USA, and other places, similar, though less serious disasters have occurred.

The effects of air pollution, and particularly the spectacular episodes mentioned above, have naturally aroused greatest alarm among the public and have prompted governments, particularly in highly industrialized countries, to develop and execute large-scale air pollution control programs. The subject of air pollution control is a specialized field, discussion of which is beyond the scope of the present text. For the sake of completeness of our consideration of atmospheric trace elements, we will therefore limit ourselves to a discussion of the production of artificial trace elements and the local atmospheric changes related with them.

ORIGIN OF AIR-POLLUTING TRACE ELEMENTS

Since the motor vehicle, burning of fossil fuels and industry are mainly responsible for the emission of air pollutants, these sources and their ground level concentrations will be discussed briefly.

The Motor Vehicle

The exhaust gases of the motor vehicle represent a very important and in some places, especially cities, the preponderant part of air pollution. The motor vehicle is propelled by the internal combustion engine. As a rule, this engine is equipped with several cylinders with moving pistons in which a gasoline and air mixture is alternately compressed and then explosively burnt. For complete combustion of the fuel, which consists of hydrocarbons, about 14.5 kg air are needed per kg of gasoline. Technically speaking, however, acceptable operation of conventional internal combustion engines under various operating conditions can only be obtained with a certain excess of fuel. As a result, because of the oxygen deficit, the fuel is burnt only partially and carbon monoxide and hydrocarbons are emitted with the exhaust gases.

The most important product of this partial oxidation is carbon monoxide, of which 3-10% is present in the exhaust gases. In addition, there are not only unconverted fuel components but also incompletely oxidized hydrocarbons, for example, aldehydes, esters, ketones, organic acids. The temperatures generated during explosion are sufficient to oxidize a small fraction of the atmospheric nitrogen and to induce cracking processes and synthesis reactions of the fuel constituents. This introduces nitrogen oxides and a number of gaseous hydrocarbons of the paraffin and olefin series and higher condensed, particularly polycylic, aromatics into the exhaust gases. Finally, aerosols are emitted, which consist primarily of condensed hydrocarbons and lead, the antiknock additive to gasoline. Additional ground-level sources of unburnt hydrocarbons are found in fuel evaporation in the gas tank and carburetor and in piston rings with a poor seat.

Table XVI gives a summary of the trace element content in motor vehicle emissions. The range of variation shown covers the various operating conditions of the engine (idling, acceleration, cruising and braking).

The emission, referred to the traveled kilometer, furnishes a criterion for the contribution of the motor vehicle to the atmospheric trace element content and is therefore of decisive importance for clean air problems. Tests performed in the US show that the exhaust gas emissions of incomplete combustion products are related to the mean speed of traffic flow. This situation is illustrated in Figure 62 by the example of studies conducted in Los Angeles. The figure reveals a 70-80% reduction of the carbon monoxide and hydrocarbon emission with a speed increase of

Table XVI

Composition of Motor Vehicle Emissions*

Constituent	*Content in %*
Nitrogen oxides, NO and NO_2	0- 0.4
Carbon dioxide, CO_2	6.5-13
Water, H_2O	7-11
Oxygen, O_2	0.1- 2
Carbon monoxide, CO	1-10
Hydrocarbons	0.02- 0.8

Aerosol emission: 0.2-3.2 mg per g of fuel consumption.

*See also VDI-Guideline No. 2282.

traffic flow from 15 to 90 km/h—from dense city traffic to expressway speeds (Ludwig, 1970). This necessarily leads to the conclusion that a considerable reduction of local vehicle emissions would be realized by the construction of freeways in the cities.

Occasionally, particularly on highly traveled streets and intersections of inner cities, vehicle carbon monoxide emissions lead to atmospheric CO concentrations (up to about 100 ppm) far above the maximum values specified with consideration of public health and welfare. In the US, for example, these maxima are set at 35 ppm as one-hour and at 9 ppm as eight-hour mean values.

Under the influence of intense solar radiation, photochemical reactions are initiated between the hydrocarbons, nitrogen oxides and oxygen in the atmosphere. The reaction products contain chemically highly reactive compounds, such as peroxides, organic radicals, ozone, and peroxyacetylnitrate, of which even very low concentrations of a few hundredths to a few tenths ppm can lead to very unpleasant irritation of the mucous membranes of the eyes and respiratory system, and can cause damage to vegetation and property (Haagen-Smit, Wayne, 1968). Los Angeles, with its basin topography favorable to inversions as well as intense solar radiation, was the first city to experience the formation of these photochemical products with strong haze development in the forties. This type of air pollution, which today is occasionally encountered in every large American city as well as in some European cities, is therefore frequently called "Los Angeles-type smog."

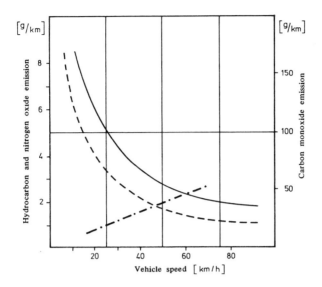

Figure 62. Mean carbon monoxide, hydrocarbon and nitrogen oxide emissions per kilometer as a function of the speed of traffic flow in Los Angeles (according to data of Ludwig, 1970). —— Hydrocarbons, - - - carbon monoxide, - - - - - nitrogen oxide.

Use of Fossil Fuels

Since its beginnings, human civilization has been characterized by the utilization of fire. In the beginning, fire served mainly for domestic purposes, such as cooking, heating and illumination. With the industrial evolution and the development of the steam engine, the energy demand multiplied with a rapid increase in the consumption of fossil fuels. Even today, we still attain our personal and industrial energy needs primarily by the combustion of coal and oil as well as to a smaller degree by natural gas. Therefore, until most recently and in a large part of the world even today, the pollutants produced by fossil fuels have continued to be the main contributors to air pollution caused by human activity.

The air pollutants produced during the combustion of coal and emitted with its waste gases fall into two groups—materials attributable to the fuel composition, for example, sulfur dioxide and flyash, and materials that originate from incomplete combustion, such as flue gases and soot.

Coal usually contains a small percentage of incombustible minerals, particularly silicates, clay, iron and calcium oxides. During coal combustion, these constituents form residues in the heating plant in the form of ash and clinker. A part, however, is entrained with the waste gases and introduced into the atmosphere in the form of flyash, unless the latter is removed from the combustion gases by precipitation. The ash particles are relatively large, having a mean diameter of about 15 microns (anonymous, 1967).

Energy generation with coal combustion is based mainly on the oxidation of carbon into carbon dioxide, which reaches the atmosphere with the waste gases. The slow increase of the atmospheric carbon dioxide content has been attributed to the worldwide CO_2 emission of about 10^{10} tons per year (see Chapter 2, page 8). Compared to the large quantities of harmless constituents of gaseous effluents—carbon dioxide, water, nitrogen and oxygen—the other air-polluting gases—sulfur dioxide, sulfur trioxide, nitrogen oxide, carbon monoxide, aldehydes and other hydrocarbons—quantitatively represent only a small fraction.

Depending on origin, coal as well as fuel oil contains about 0.5-4% sulfur in the form of pyrites and organic compounds. During combustion this sulfur is oxidized into sulfur dioxide, SO_2, and a small percentage of sulfur trioxide, SO_3, and is emitted into the atmosphere together with the flue gases. Sulfur trioxide combines with atmospheric humidity to form the extremely corrosive sulfuric acid. If air is heated in a hot flame, a fraction of the nitrogen combines with oxygen to form nitrogen oxides. With slow cooling this process is reversible and the nitrogen oxides dissociate into their components. In thermal power plants, however, the heat generated is removed as rapidly as possible from the combustion gases in order to utilize it profitably. Under these conditions the nitrogen oxides remain partially bound in the waste gases and are emitted to the atmosphere in concentrations of a few hundred ppm.

For economy reasons, efforts are made to utilize the fuel as completely as possible in thermal power plants. In such plants, emission of incomplete combustion products is therefore usually limited to a minimum. In contrast, the coal furnaces used for residential heating are often wasteful and contribute considerably to air pollution during the heating period by smoke and soot production, particularly in cities.

Table XVII, a summary of gaseous emissions of different coal consumers, as expected shows that sulfur dioxide and nitrogen

Table XVII

Emissions of Air-Polluting Gases in Coal Combustion*

	Electrical Power Plants	Industrial Power Plants	Residential Heating
Aldehydes	0.003	0.003	0.003
Carbon monoxide	0.25	1.5	25
Hydrocarbons	0.1	0.5	5
Nitrogen oxides	10	10	4
Sulfur dioxide**	40	40	40

*According to Rossano, Jr., 1969. Data expressed in kg per ton of burnt coal.
**Data based on a coal sulfur content of 2%.

oxide formation is approximately identical for all plants. However, compared to large-scale installations, residential combustion produces approximately 10-100 times more incompletely burnt gases, i.e., carbon monoxide and hydrocarbons.

The SO_2 and SO_3 emissions from coal and crude oil combustion lead to considerable damage of vegetation and property. In many cases efforts are therefore underway to limit sulfur oxide concentrations to an acceptable ground-level concentration. At the moment, attempts are being made to realize this objective primarily by the construction of high smokestacks. It is to be hoped that the methods for coal and crude oil desulfurization and elimination of sulfur oxides from gaseous effluents that are presently under development will lead to a further reduction in the future in spite of increasing energy demands.

The conversion of residential heating from coal to coke heating in slow-burning furnaces and the change to oil and gas heating have already led to a noticeable reduction of smoke and dust emissions. A desirable goal from the standpoint of clean air would be the complete elimination of residential heating as a source of air pollution by conversion to electrical heating.

Industrial Air Pollutants

Undoubtedly, industry in its concentrated locations, for example, in the Ruhr region, contributes predominantly to air pollution. Depending on the sector of industry or the production processes, a tremendous number of diverse materials and chemical compounds are introduced into the atmosphere.

All industrial plants are equipped with heating systems and frequently also with thermal power plants, which introduce primarily sulfur dioxide and nitrogen oxides into the atmosphere. Foundries with their production of dust, metal oxide fume, soot and sulfur dioxide (and occasionally also fluorides) and cement plants with their dust generation represent the main contributors to industrial air pollution. The emissions of the chemical industry are to be attributed to the release of impurities of raw materials, losses of operating materials, and incomplete recovery of end products, among other things. The inorganic chemicals industry is represented by various gaseous effluents, for example, SO_2, NO, NO_2, H_2SO_4, HCl, and FH, depending on the nature of their production. Organic chemical plants (such as oil refineries and paper pulp production) are represented primarily by sometimes malodorous organic compounds.

A reduction of industrial emissions can be achieved technically in many ways. High smokestacks are utilized in order to transport undesirable gaseous effluents to such altitudes that the pollutants will be diluted sufficiently by meteorological influences (eddy diffusion, wind, etc.) before they reach the populated surface of the earth, so that presumably they will not cause damage there. Combustible effluents can either be used further as heating sources or can be rendered harmless by bleeding off in torches or by catalytic oxidation. In many cases the pollutants can be extensively removed from the effluents by absorption methods or scrubbing. From the economy standpoint, these processes are advantageous particularly when the solvents or adsorbents are recoverable and the pollutant removed has economic value. Methods for the removal of solids have probably advanced to the greatest degree. Depending on the composition of the gaseous effluents and the particles suspended in them, cyclones, filter columns, electrostatic precipitators and scrubbers can be utilized for their removal.

LOCAL AIR POLLUTION

The anthropogenic contributor to the worldwide trace element concentration has been discussed in the preceding chapters to the extent to which it is known today. Beyond this, air pollutants are of considerable importance in the closer vicinity of their emission sources, since under certain meteorological conditions they may attain concentrations that are orders of magnitude greater than their global values. Table XVIII illustrates this fact for various pollutants. Measurements in Chicago give an idea of the mean annual concentrations and highest diurnal means that were observed in this industrial city in 1964. As expected, these values considerably exceed the mean tropospheric values on the last line.

Table XVIII

Comparison of the Concentrations of Air Pollutants in Chicago, 1964*

	SO_2 ppm	CO ppm	NO_2 ppm	Aerosols $\mu g/m^3$
Chicago				
24-h maximum	0.68	27	0.15	–
Annual mean	0.18	12	0.05	177
Tropospheric mean	0-0.02	0.01-0.02	0.001-0.005	20-60

*According to Tebbens, 1968, with mean tropospheric values (Junge, 1963).

This locally increased concentration of air pollution is causally related with the repeatedly mentioned injury to health and property and consequently is of particular interest for air pollution control. Furthermore, certain changes in local climate, for example, frequency of fog, increased precipitation, and impaired visibility, can be attributed to increased air pollution.

The local pollutant concentration is essentially determined by two factors, *i.e.*, the emission rate of gaseous effluent sources and their dilution in the environment by wind and turbulent exchange. The source extension is important for the effluent dilution between emission site and the point where the gaseous effluents can cause damage. In accordance with their configuration, three different types of sources are therefore distinguished, point sources, line sources and area sources. Smokestacks standing alone are an example of the first, uniformly and heavily traveled long-distance highways illustrate the second, while urban zones with their innumerable single sources (residential chimneys, motor vehicles), which are distributed more or less uniformly over an extended area, are an example of the third. Gaseous effluent dilution increases with wind speed. This can be easily illustrated by the following example.

We assume that the chimney emits one unit of sulfur dioxide into the air, which passes the chimney with a velocity of one meter per second. If the wind speed increases to two meters per second, the air volume passing the chimney per unit time is doubled and the resulting sulfur dioxide content is divided in half. In the case of an area source, for example, an urban region, the conditions are somewhat more complicated. However, here too, wind speed plays an important role in the reduction of air pollution.

The gustiness of the wind, which becomes manifest in a continuous fluctuation of wind direction and speed, has the result that a smoke plume is continuously broadened with increasing distance from the chimney so that its gaseous effluent concentration constantly decreases. In the case of a continuous line source, this dilution is not possible and for extended area sources it also plays only a minor role.

Another factor that decisively controls atmospheric pollutant concentrations is vertical exchange, which is closely related with the atmospheric temperature stratification (see also page 82). Good vertical exchange is limited to the air layer between the surface of the earth and the tropopause—the barrier between troposphere and stratosphere. The availability of this entire layer for the dilution of air pollutants, however, depends highly on the temperature trend in this region of the atmosphere. If we neglect phase changes of water vapor, good mixing requires a temperature decrease with altitude of more than $10°C$ per kilometer (dry adiabatic temperature gradient). With such a temperature structure, the specific gravity of the air increases with altitude. This instability leads to the rise of bubbles of lighter air and the descent of heavier air, thus producing vigorous vertical exchange. With a decreasing temperature gradient, vertical exchange diminishes and in the case of a temperature increase with altitude (inversion), when a cold air layer is covered by lighter warmer air, it is practically arrested.

Ordinarily the temperature profile is subject to pronounced diurnal fluctuations. Shortly after sunrise, heating of the surface of the earth leads to rapid heating of the air and thus to good convection near ground level. The thickness of this mixing layer is determined by the intensity of solar radiation and the radiation properties of the soil, and in the summer this usually amounts to several hundred up to about two thousand meters. The conditions change drastically on clear or slightly cloudy nights with little air motion. After sundown, the surface of the earth cools rapidly by radiation loss, which results in cooling of the overlying air blanket. This results in the formation of an inversion layer originating from the ground, the thickness of which usually increases in the course of the night. Air pollutants emitted into this layer consequently remain confined to a very shallow layer (about 50-100 m) near the emission level and therefore attain correspondingly high concentrations. Long fall and winter nights favor the formation and preservation of these ground inversions. Episodes of extremely high air pollution levels occur primarily during these seasons.

Two other processes influence vertical eddy diffusion to a significant degree. The thermal capacity of urban zones, and, to a lesser degree, the heat released by fuel consumption, modify the urban temperature profile. This effect is most pronounced at night. Heat stored in buildings and on roads during the day is slowly released to the atmosphere at night and prevents the development of ground inversions. The urban influence frequently extends to an altitude of several hundred meters, so that the emissions are distributed at least over a layer of this thickness.

Large scale weather systems can influence the temperature profile on a far broader basis, over thousands of square kilometers. In moving low-pressure zones (cyclones), the wind connected with strong atmospheric pressure changes and the inflow of air into the storm center produce good ventilation and vertical mixing. On the other hand, ventilation and vertical mixing are often greatly reduced under high-pressure weather conditions (anticyclones). The small pressure differences of anticyclones produce only weak wind and consequently poor ventilation in air-polluted regions. Near ground level, air continuously flows away from the high pressure zone. For reasons of continuity, this air must be replaced by air from higher layers. The descending air undergoes adiabatic heating and finally forms an inversion aloft which prevents the transport of air pollutants beyond this barrier. In unfavorable cases, in stationary high-pressure zones, the barrier layer may slowly drop to the surface and may thus concentrate the air pollutants in a very thin ground layer for the duration of these weather conditions. It is the latter that, among other things, are responsible for the beautiful weather of "Indian summer" as well as for the well-known air pollution disasters in London, the Maastal, Donora, and elsewhere.

Meteorology essentially has the difficult, although challenging, task to establish the relationship between the atmospheric pollutant concentration and emissions from known sources under various meteorological conditions. Although considerable advances have been made in recent years in the field of air pollution forecasting, numerous questions remain unresolved because of the complexity of the problem.

Today, the best results of forecasting air pollution under various meteorological conditions can be accomplished for isolated point sources, provided the local topography is not too complex. This procedure, which starts with the source strength, calculates the rise and dispersion of the smoke plume from the parameters of horizontal and vertical exchange as well as wind speed (see Strom, 1968). This then allows a determination of the pollutant concentrations for every downwind location.

A treatment of air pollution in urban and industrial regions becomes considerably more difficult. The large number and diversity of emission sources together with the frequently complex terrain make an individual treatment impossible. One of the main sources of our cities, the motor vehicle, does not even remain stationary. A further complication and one of the most interesting aspects of the entire complex of problems is that meteorological influences are subject to marked fluctuations over short periods of time, while emissions may be considered to be relatively constant. However, this picture is reversed over longer periods of time, in the order of years. The mean meteorological conditions become fairly constant over such periods, while emissions vary considerably as a result of changes in industrialization, population density, standard of living and, hopefully, stricter emission controls.

Different methods must be used depending upon the problem involved: if one considers short periods of time and asks, *e.g.* for the statistical distribution of air pollutant concentrations, the behavior of different pollutants in the atmosphere, or the influence of air pollution on weather, one often starts with actual measurements of the atmospheric pollution concentration without a specific knowledge of sources and their strength, and one attempts to correlate these measurements statistically with the meteorological parameters. This technique has furnished interesting information, for example, with regard to the atmospheric lifetime of pollutants, annual fluctuations of air pollution, the role of sunlight for photochemical smog formation and the reduction of solar radiation in cities, to mention just a few.

Questions concerning the origin of a specific pollutant, the efficiency of new emission control systems, or the change of the pollutant content and its composition due to growth require a model based on sources. Instead of starting with innumerable single sources, attempts are generally made to combine the emission sources into groups (industry, traffic, residential heating) and on the basis of principal air pollutants (sulfur dioxide, carbon monoxide), and to classify them according to geographical location. One of the most promising advances of recent years is the development of mathematical models that incorporate such information and, together with the meteorological parameters, make it possible to calculate the concentration field of urban air pollution.

The application of numerical mathematical models that can be solved by computers appears to be highly promising for a quantitative description of air pollution. It is to be hoped that our continuously improving understanding of the meteorology of cities and the dispersion and removal of pollutants will make it possible to predict air pollution concentrations routinely and thus under unfavorable conditions to supply control authorities with the necessary information to take prompt measures and thus prevent catastrophic episodes in the future.

Appendix

SETTLING VELOCITY OF AEROSOL PARTICLES

(Appendix to Chapter 3, pages 32 and 45).

If the expression $m \cdot g$ - m = particle mass, g = acceleration due to gravity, is introduced into Equation (13) for K, the falling speed v_g of the respective particle is given by:

$$v_g = m\, g\, C_c / 6\, \pi\, \eta\, r \tag{50}$$

$$C_c = \text{``Cunningham correction''} = 1 + \frac{L}{r}\, (A + Be^{-Cr/L})$$

[A, B and C are empirical dimensionless constants—compare Equation (13)]

$$L = \text{mean free pathlength.}$$

If we set $m = \frac{4}{3} r^3\, \pi\rho$, where ρ is the particle density, v_g becomes:

$$v_g = 2/9\, r^2\, \rho\, g\, C_c / \eta \tag{51}$$

F. Kasten (1968) has published detailed tables of the settling velocity for eight particle radii between 0.003 and 10 μm in different altitudes of the atmosphere between 0 and 80 km on the basis of the values of "1962 US Standard Atmosphere" (A. E. Cole, A. Court and A. J. Kantor, 1965). Table XIX shows an excerpt from the tabulation of F. Kasten.

Table XIX

Settling Velocities in cm/sec of Spherical Particles of 1 g/cm³ Density at Different Altitudes of the Model Atmosphere "1962 US Standard Atmosphere"*

Altitude in km	Particle radius in µm					
	0.003	0.01	0.03	0.1	1	10
0	$4.1150 \cdot 10^{-6}$	$1.4283 \cdot 10^{-5}$	$4.8096 \cdot 10^{-5}$	$2.3182 \cdot 10^{-4}$	$1.3188 \cdot 10^{-2}$	$1.2280 \cdot 10^{0}$
2	5.1727	1.7825	5.8811	2.6796	1.3890	1.2754
4	6.5809	2.2537	7.3058	3.1600	1.4725	1.3264
6	8.4837	2.8901	9.2287	3.8083	1.5736	1.3848
8	$1.1096 \cdot 10^{-5}$	3.7630	$1.1864 \cdot 10^{-4}$	4.6970	1.6985	1.4509
10	1.4747	4.9831	1.5544	5.9377	1.8575	1.5269
15	3.1959	$1.0721 \cdot 10^{-4}$	3.2758	$1.1646 \cdot 10^{-3}$	2.3540	1.6052
20	6.9792	2.3332	7.0582	2.4227	3.4878	1.6972
25	$1.5160 \cdot 10^{-4}$	5.0598	$1.5237 \cdot 10^{-3}$	5.1462	6.0997	1.8725
30	3.2344	$1.0788 \cdot 10^{-3}$	3.2420	$1.0872 \cdot 10^{-2}$	$1.1757 \cdot 10^{-1}$	2.3077
40	$1.3688 \cdot 10^{-3}$	4.5633	$1.3695 \cdot 10^{-2}$	4.5710	4.6488	5.5071
50	4.9889	$1.6630 \cdot 10^{-2}$	4.9896	$1.6638 \cdot 10^{-1}$	$1.6710 \cdot 10^{0}$	$1.7453 \cdot 10^{1}$
60	$1.7463 \cdot 10^{-2}$	5.8210	$1.7463 \cdot 10^{-1}$	5.8217	5.8293	5.9050
70	6.8892	$2.2964 \cdot 10^{-1}$	6.8893	$2.2965 \cdot 10^{0}$	$2.2974 \cdot 10^{1}$	$2.3058 \cdot 10^{2}$
80	$3.5574 \cdot 10^{-1}$	$1.1858 \cdot 10^{0}$	$3.5574 \cdot 10^{0}$	$1.1858 \cdot 10^{1}$	$1.1859 \cdot 10^{2}$	$1.1859 \cdot 10^{3}$

*According to F. Kasten, 1968.

Literature

Ackermann, P., Kondensationskernzahlung in Payerne 1953. (Counting of condensation nuclei in Payerne 1953), Geofisica pura e applicata *29*, 168-177 (1954).

Aitken, J., On a simple pocket dust-counter, Proc. Royal Soc. Edinburgh *18*, (s. Coll. Scient. Papers, Nr. 18) (1890/91).

Aitken, J., On the number of dust particles in the atmosphere, Trans. Royal Soc. Edinburgh *35*, 1-19, 1887/88. Sowie zahlreiche andere Arbeiten (s. Aitken Collected Scientific Papers, Cambridge Univ. Press) (1923).

Ångström, A., On the atmospheric transmission of sun radiation and on dust in the air (I), Geografiska Annaler, Årg. XI, 156-166 (1929).

Ångström, A., On the atmospheric transmission (II), Geografiska Annaler, Årg. XII, 130-159 (1930).

Anonymous, Criteria for the application of dust collectors for coal fired boilers, Ind. Gas Cleaning Institute, New York 1967.

Arnold, P. W., Losses of nitrous oxide from soil, Journ. Soil Sciences *5*, 116-128 (1954).

Berichte des Deutschen Wetterdienstes *14*, Nr. 100 (1965).

Bhandari, N., Rama, Atmospheric circulation from observations of Sodium-22 and other short lived natural radioactivities, JGR *68* (1963).

Bhandari, N., Lal, D., Rama, Stratospheric circulation studies based on natural and artificial radioactive tracer elements, Tellus *18*, 391-405 (1966).

Bigg, E. G., The detection of atmospheric dust and temperature inversions by twilight scattering, Journ. Meteorology *13*, 262-268 (1956).

Birot, A., Fontan, J., Adranger, B., Blanc, D., Bouville, A., Measurement of radon concentration in the troposphere up to 5000 meters. Paper presented at the Journées d'Electronique de Toulouse, p. 1-7 (March 1968).

Bolin, B., Transfer and circulation of radioactivity in the atmosphere. In: Kernstrahlung in der Geophysik, (H. Israel, A. Krebs), Springer 1962.

Bolin, B., Note on the exchange of iodine between the atmosphere, land and sea, Internat. Journ. of Air Pollution *2*, 127-131 (1959).

Bradley, W. E., Pearson, J. E., Measurement of the vertical distribution of radon in the lower atmosphere, JGR 75, 5890-5894 (1970).

Bullrich, K., Streulichtmessungen in Dunst und Nebel. (Scattered light measurements in haze and fog), Meteorol. Rundschau, *13*, 21-29 (1960).

Burton, W. M., Stewart, N. G., Use of long lived natural radioactivity as an atmospheric tracer, Nature *186*, 584 (1960).

Callendar, G. S., On the amount of carbon dioxide in the atmosphere, Tellus *10*, 243-248 (1958).

Chagnon, C. W., Junge, C. E., The vertical distribution of submicron particles in the stratosphere, Journ. Meteorology *18*, 746-752 (1961).

Chambers, L. A., Milton, J. F. and Cholak, C. E., A comparison of particulate loadings in the atmosphere of certain American cities. Paper presented at Third National Air Pollution Symposium, Pasadena (1955).

Cole, A. E., Court, A., and Kantor, A. J., Model atmospheres, Handbook of Geophysics and Space Environments, Bedford, Mass., U.S. Air Force Cambridge Research Laboratories 2/1-2/22 (1965).

Coulier, M., Note sur une nouvelle propriété de l'air. (On a new property of the air), Journ. de Pharmac. et de Chimie (4) *22*, 165-173, 254-255 (1875).

Craig, H., The natural distribution of radiocarbon and the exchange time of carbon dioxide between the atmosphere and the sea, Tellus *9*, 1-17 (1957).

Crozier, W. D., Black magnetic spherules in sediments, Journ. Geophys. Res. *65*, 2971-2974 (1960).

Davies, C. N., "Aerosol Science," Academic Press, London and New York, 468 (1966).

de Bary, E., Bullrich, K., Über den Anteil der Rayleighstreuung und den Einfluß der Aerosolgrößenverteilung auf die Extinktion und spektrale Intensität der Streustrahlung eines Luftvolumens. (On the contribution of Rayleigh scattering and the influence of aerosol size distribution on the absorbance and spectral intensity of the scattered radiation of an air volume), Archiv f. Meteorol., Geophys. u. Bioklim. (B) (1962).

Dolezalek, H., The atmospheric electric fog effect, Review of Geophysics *1*, 231-282 (1963).

Eriksson, E., The yearly circulation of chloride and sulfur in nature; meteorological, geochemical and pedological implications, Part I, Tellus *11*, 375-403, Part II, Tellus *12*, 63-109 (1959).

Facy, L., La capture des noyaux de condensation par chocs moléculaires au cours des processus de condensation. (Capture of condensation nuclei by molecular collisions during condensation processes), Archiv Meteor., Geophys., Bioklim. (A) 8, 229-236 (1955).

Facy, L., Sur le déplacement des particules d'aérosols au cours des processus de diffusion moléculaires. (On the displacement of aerosol particles during molecular diffusion processes), Compt. Rend. Acad. Sci. Paris *246*, 102-104. See ibid, *246*, 3161-3164, 1958, (On a capture mechanism of aerosol particles by a droplet in the condensation-evaporation process) (1958).

Facy, L., Radioactive precipitation and fall out. In: H. Israël und A. Krebs, Kernstrahlung in der Geophysik—Nuclear Radiation in Geophysics, 203-240 (1962).

Feely, H. W., Seitz, H., Use of lead-210 as a tracer of transport processes in the stratosphere, JGR 75, 2885-2894 (1970).

Fenn, R. W., Measurements of the concentration and size distribution of particulates in the arctic air of Greenland, USASRDL (US Army Signal Research and Development Laboratory), Techn. Rep. 2097 (1960).

Friedlander, S. K., Similarity considerations for the particle size spectrum of a coagulating, sedimenting aerosol, Journ. Meteorol. *17*, 479-483 (1960).

Georgii, H. W., Untersuchungen über atmosphärische Spurenstoffe und ihre Bedeutung für die Chemie der Niederschlage. (Studies on atmospheric trace elements and their importance for the chemistry of precipitation), Geofisica pura e applicata *47*, 155-171 (1960).

Georgii, H. W., Untersuchungen über Ausregnen und Auswaschen atmosphärischer Spurenstoffe durch Wolken und Niederschlag. (Studies on rainout and washout of atmospheric trace elements by clouds and precipitation), Berichte d. Dt. Wetterd. *14*, Nr. 100, 23 (1965).

Georgii, H. W., Die Verteilung von Spurengasen in reiner Luft. In: "Atmosphärische Spurenstoffe und ihre Bedeutung für den Menschen." (Distribution of trace gases in clean air), In: "Atmospheric Trace Elements and Their Importance for Man." [Symposium at the Physiological Weather Station of St.-Moritz-Bad (Engadin), 18-19 June 1966], 14-20, Birkhäuser Verlag, Basel (Experientia Supplement 13) (1967).

Gish, O. H., Sherman, K. L., Electrical conductivity of air to an altitude of 22 kilometers, Nat. Geogr. Soc., Techn. Papers, Stratospheric Series, Nr. 2, Washington 1936.

Glawion, H., Staub und Staubfälle in Arosa, (Dust and dust fall-out in Arosa), Beitr. z. Phys. d. fr. Atm. *25*, 1-43 (1938).

Goetz, A., Ursprung, Verhalten und Bestimmung der Submikronen-Aerosole des Smogs, (Origin, properties and determination of the submicron aerosols of smog), Staub *20*, 303-308 (1960).

Goetz, A., Preining, O., The aerosol spectrometer and its application to nuclear condensation studies. In: "Physics of Precipitation," Am. Geoph. Union, Monogr. *5*, 164-182, Washington, D.C. (1960).

Goldsmith, P., Delafield, H. J., and Cox, L. C., Measurement of the deposition of submicron particles in gradients of vapour pressure and the efficiency of this mechanism in the capture of particulate matter by cloud droplets in nature, Geofisica pura e applicata *50*, 278-280 (1961).

Goldsmith, P., May, F. G., Diffusio- and Thermophoresis in water vapour systems. In: C. N. Davies, Aerosol Science, 163-194 (1966).

Goody, R. M., Walshaw, C. D., The origin of atmospheric nitrous oxide, Quart. Journ. Royal Meteorol. Soc. *79*, 496-500 (1953).

Gotsch, G., Untersuchungen zum Problem der Aktivität kleiner Kondensationskerne, (Studies of the problem of the activity of smog condensation nuclei), Archiv. f. Meteorol., Geophys. u. Bioklimat. (A) *13*, 73-116 (1962).

Grozier, W. D., Direct measurement of radon-220 (thoron) exhalation from the ground, JGR *74*, 4199-4205 (1969).

Guedalia, D., Laurent, J. L., Fontan, J., Blanc, D., Druilhet, A., A study of Rn-220 emanation from soils, JGR *75*, 357-369 (1970).

Haagen-Smit, A. J., Wayne, G., Atmospheric reactions and scavenging processes, in: Air Pollution, Vol. I (A. C. Stern Herausgeber) Academic Press, New York 1968.

Haxel, O., Schumann, G., Erzeugung radioaktiver Kernarten durch die kosmische Strahlung, (Production of radionuclide species by cosmic radiation), In: Kernstrahlung in der Geophysik (H. Israel, A. Krebs), Springer 1962.

Herpertz, E., Condensation nuclei and atmospheric conductivity at Jungfraujoch (Switzerland), Techn. Note No. 9, Contract AF *61* (514)-640, Aachen 1957.

Hodges, P. W., Sampling dust from the stratosphere, Smithsonian Contrib. to Astrophysics *5*, 145-152 (1961).

Hodges, P. W., Wright, F. W., The space density of atmospheric dust in the altitude range 50000 to 90000 feet., Smithsonian Contrib. to Astrophysics *5*, 231-238 (1962).

Horbert, M., Untersuchungen zur atmosphärischen Turbulenz mittels Radon-220 als Tracer, (Studies of atmospheric turbulence by Radon-220 as tracer), Dissertation RWTH-Aachen 1969.

Hunter, H. F., Ballou, N. E., Nucleonics *9*, C2-C5 (1951).

Hutchinson, G. E., The biochemistry of the terrestrial atmosphere. In: "The earth as a planet," (Edited by G. P. Kuiper), pp. 371-433, Univ. of Chicago Press, Chicago, Ill. 1954.

Israel, G. W., Der Thorongehalt in der bodennahen Atmosphäre, (Thoron content of the atmosphere near ground level), Dissertation der Technischen Hochschule Aachen 1965.

Israel, H., In: Compendium of Meteorology (T. F. Malone, Ed.), Boston, 155-161 (1951).

Israel, H., Die Kondensationskerne als Bindeglied luftelektrisch-meteorologischer Zusammenhänge, (Condensation nuclei as the linking member of meteorological electrical relationships), Geofisica pura e applicata *31*, 162-168 (1955), (s. auch ebenda *36*, 182-200, 1957).

Israel, H., Condensation nuclei as connecting links for meteorological electrical relations, I and II, Techn. No. 1 and 11, Contract AF 61 (514)-640, Aachen 1955/56.

Israel, H., Kondensationskerne im Rahmen der Luftelektrizitat, (Condensation nuclei as a part of atmospheric electricity), Geofisica pura e applicata *36*, 182-200 (1957).

Israel, H., Atmospheric electric and meteorological investigations in high mountain ranges, ("Alps-Project 1954-1957"), Final Rep. Contract AF 61 (514)-640, Aachen 1957a.

Israel, H., Atmosphärische Elektrizität, I. Band (370 pp.), Leipzig 1957b.

Israel, H., Atmosphärische Elektrizität, II. Band (503 pp.), Leipzig 1961.

Israel, H., Die natürliche und künstliche Radioaktivität in der Atmosphäre, (Natural and artificial radioactivity in the atmosphere), In: Kernstrahlung in der Geophysik, (H. Israel and A. Krebs), Springer 1962.

Israel, H., Radioaktivität der Atmosphäre, (Radioactivity of the atmosphere), In: Atmosphärische Spurenstoffe und ihre Bedeutung für den Menschen, Birkhauser Verlag 1967.

Israel, H., Dolezalek, H., On the application of Facy-forces to atmospheric ions, (Paper before the symposium on Atmospheric and Space Electricity during the meeting of the International Union of Geodesy and Geophysics (IUGG) in Berkeley, Calif. 1963. Abstract in Publication No. 13 IAMAP (Intern. Association of Meteorol. and Atm. Physics), Toronto, 152 (1963).

Israel, H., Horbert, M., de la Riva, C., The Thoron Content of the Atmosphere and its Relation to the Exchange Conditions, Final Technical Report for US Dpt. of the Army. Contract DA-91-591-EUC-3761, Aachen 1966.

Israel, H., Israel, G. W., Neues Verfahren zur Abscheidung radioaktiver Aerosole, (New method for the precipitation of radioactive aerosols), Naturwissenschaften *49*, 373 (1962).

Israel, H., Nix, N., Thermodynamic processes in the condensation nuclei counter, Journ. de Recherches Atmosphériques, Vol. II, *2e année*, No. 2-3, 185-187 (see also ZS f. Geophys. *32*, 175-177) (1966).

Israel, H., Nix, N., Ein neues Verfahren zur Untersuchung von Kondensation und Verdampfung an Einzelteilchen von Kleinaerosolen, (A new method for the study of condensation and evaporation of individual particles of small aerosols), Zs. f. Geophys. *35*, 207-209 (1969).

Israel, H., Schulz, L., Über die Größenverteilung der atmosphärischen Ionen, (On the size distribution of atmospheric ions), Meteorol. Zeitschr. *49*, 226-233 (1932).

Israel, H., Stiller, S., Climatological aspects of the natural radioactivity, Zs. f. Geophysik *29*, 51-56 (1963).

Jacobi, W., Natural atmospheric radioactivity, Bericht des Hahn-Meitner-Instituts für Kernforschung, (Report of the Hahn-Meitner Institute of Nuclear Research), Berlin-Wannsee 1960.

Jacobi, W., André, K., The vertical distribution of radon-220, radon-222, and their decay products in the atmosphere, JGR *68*, 3799 (1963).

Jaufmann, H., Deutsches Meteorol. Jahrbuch Bayern, (German Meteorological Yearbook Bavaria), 29 (1907).

Junge, C. E., Zur Frage der Kernwirksamkeit des Staubes, (On the question of the particulate efficiency of dust), Meteorol. Zeitschr. *53*, 186-188 (1936).

Junge, C. E., Nuclei of atmospheric condensation. In: "Compendium of Meteorology," 182-191, Boston 1951.

Junge, C. E., Die Rolle der Aerosole und der gasförmigen Beimengungen der Luft im Spurenstoffhaushalt der Troposphäre, (The role of aerosols and gaseous admixtures of air in the trace element balance of the atmosphere), Tellus *5*, 1-26 (1953).

Junge, C. E., The chemical composition of atmospheric aerosols. In: Measurements at Round Hill Field Station, June-July 1953, Journ. of Meteorol. *11*, 323-333 (1954).

Junge, C. E., Recent investigations in air chemistry, Tellus *8*, 127-139 (1956).

Junge, C. E., Remarks about the size distribution of natural aerosols. In: "Artificial Stimulation of Rain," Proc. 1st Conf. Physics Cloud and Precipitation Particles (edited by H. Weickmann and W. Smith), Pergamon Press, 3-17 (1957).

Junge, C. E., Vertical profiles of condensation nuclei in the stratosphere, Journ. Meteorol. *18*, 501-509 (1961).

Junge, C. E., Radioaktive Aerosole, (Radioactive aerosols), In: Kernstrahlung in der Geophysik, (H. Israel, A. Krebs), Springer 1962.

Junge, C. E., Air chemistry and radioactivity, (International Geophysics Series, Vol. 4) Academic Press, New York, 382 (1963).

Junge, C. E., Considerations about the ozone budget. Cited according to Junge, C. E., pp. 56 ff. (1963).

Junge, C. E., Übersicht über die wesentlichen Eigenschaften der natürlichen Aerosole, (Summary of the significant properties of natural aerosols), Experientia, Supplementum *13*, 9-13 (1967).

Junge, C. E., Manson, J. E., Stratospheric aerosol studies, Journ. Geophys. Res. *66*, 2163-2182 (1961).

Junge, C. E., Werby, R. T., The concentration of chloride, sodium, potassium, calcium and sulfate in rain water over the United States, Journ. of Meteorol. *15*, 417-425 (1958).

Kalkstein, M. I., Movement of material from high altitude deduced from tracer observations, JGR *68*, 3835 (1963).

Kasten, F., Der Einfluß der Aerosol-Größenverteilung und ihrer Änderung mit der relativen Feuchte auf die Sichtweite,

(Influence of the aerosol size distribution and its variation with relative humidity on visibility), Beitr. Phys. d. Atm. *41*, 33-51 (1968).

Kasten, F., Falling speed of aerosol particles, Journ. of Applied Meteorol. *7*, 944-947 (1968).

Keefe, D., Nolan, P. J., and Rich, T. A., Charge equilibrium in aerosols according to the Boltzmann law, Proc. Royal Irish Acad. (A) *60*, 27-45 (1959).

Kientzler, C. F., Arons, A. B., Blanchard, D. C., and Woodcock, A. H., Photographic investigation of the projection of droplets by bubbles bursting at a water surface, Tellus *6*, 1-7 (1954).

Lal, D., Peters, B., Cosmic ray produced isotopes and their application to problems in geophysics, Progress in Elementary Particle and Cosmic Ray Physics *6*, 3 (1962).

Lal, D., Peters, B., Cosmic ray produced radioactivity on the earth. Handbuch der Physik, (S. Flügge) Band 46/2, Springer.

Landolt-Börnstein, Zahlenwerte und Funktionen, (Numerical tables and functions), 6th Edition, vol. 3, pp. 306 ff. (1952).

Landsberg, H., Observations of condensation nuclei in the atmosphere, Monthl. Weather Rev. *62*, 442-445 (1934).

Landsberg, H., Atmospheric condensation nuclei, Ergebn. d. kosm. Phys. III. Band, 155-252 (1938).

Langevin, P., Recombinaison et mobilites des ions dans les gaz, (Recombination and mobilities of ions in gases), Ann. Chim. Phys. *28*, 433-530 (1903).

Lenard, P., Über Elektrizitätsleitung durch freie Elektronen und Träger II: Wanderungsgeschwindigkeit kraftgetriebener Partikel in reibenden Medien, (Electrical conduction through free electrons and carriers. II: Velocity of force-driven particles in frictional media), Ann. d. Phys. *60*, 329-380 (1919).

Lenard, P., Über Elektrizitätsleitung durch freie Elektronen und Träger III: Wanderungsgeschwindigkeit kraftgetriebener Partikel in reibenden Medien, (Electrical conduction through free electrons and carriers. III: Velocity of force-driven particles in frictional media), Ann. d. Phys. *61*, 665-741 (1920).

Lettau, H., Über die Zeit- und Höhen-Abhängigkeit des Austauschkoeffizienten im Tagesgang innerhalb der Bodenschicht, (Concerning the time- and altitude-dependence of the exchange coefficient in the diurnal fluctuation within the surface layer of the earth), Gerlands Beitr. z. Geophys. *57*, 171-192 (1941); see also: Landolt-Börnstein: Numerical tables and functions,

6th Edition, vol. III: Astronomy and Geophysics, Springer 1952.

Lockhart, L. B., Patterson, R. L., Saunders, A. W., Black, R. W., Fission product radioactivity in the air along the 80th meridian (west), during 1959, JGR *65*, 3987 (1960).

Lockhart Jr., B., Patterson, R. L., Saunders, A. W., JGR *71*, 1985-1991 (9166).

Loeb, L. B., "Fundamental processes of electric discharges in gases," New York 1947.

London, J., The distribution of total ozone over the Northern Hemisphere, Sun and Work *7*, No. 2, 11-12 (1962).

Ludwig, J. H., Seminar on air pollution by motor vehicles, unpublished (1968).

Martell, E. A., J. Atmospheric Science *25*, 113 (1968).

Mason, B. J., Bursting of air bubbles at the surface of sea water, Nature 174, 470-471 (1954).

Mason, B. J., "The Physics of Clouds," Oxford University Press 1957.

Metnieks, A. L., The size spectrum of large and giant sea-salt nuclei under maritime conditions, Geophys. Bulletin, School of Cosm., Phys., Dublin *15*, 1-50 (1958).

Metnieks, A. L., Pollak, L. W., On the particle size analysis of poly-disperse aerosols using a diffusion battery and the exhaustion method, Geophys. Bulletin No. 21, School of Cosmic Physics, Dublin, 53 (1962).

Moore, D. J., Measurements of condensation muclei over the North Atlantic, Quart. Journ. Royal Met. Soc. *78*, 596-602 (1952).

Münnich, K. O., Heidelberg natural radiocarbon measurements I, Science 126, 194-199 (1957).

Münnich, K. O., Vogel, J. C., Durch Atombomben erzeugter Radiokohlenstoff in der Atmosphäre (Radioactive carbon in the atmosphere produced by nuclear weapons), Naturwissenschaften *14*, 327-329 (1958).

Neumann, H. R., Messungen des Aerosols an der Nordsee, (Measurement of the aerosols on the North Sea), Gerlands Beitr. z. Geophysik *56*, 49-91 (1940).

Nix, N., Die Kondensation und Verdampfung an Einzelteilchen von Kleinaerosolen, Staub 29, 188-191 (1969a)

Nix, N., Untersuchung zur Kondensation und Verdampfung an künstlichen und näturlichen Aerosolen, (Study on the condensation and evaporation of artificial and natural aerosols), Dissertation Aachen 1969b.

Nolan, P. J., Experiments on condensation nuclei, Proc. Royal Irish Acad. *47*, 25-38 (1941).

Nolan, P. J., Kennan, E. L., Condensation nuclei from hot platinum: Size, coagulation coefficient and charge-distribution, Proc. Royal Irish Acad. *52*, 171-190 (1949).

Oepic, E. J., Interplanetary dust and terrestrial accreation of meteoric matter, Irish Astron. Journ. *4*, 84-135 (1957).

Pascerie, R. E., Friedlander, S. K., Measurements of the particle size distribution of the atmospheric aerosol: II: Experimental results and discussion, Journ. Atm. Science *22*, 577-584 (1965).

Pearson, J. E., Radon-222, a study of its emanation from soil, source strength, and use as a tracer, Research report, Dept. of General Engineering, University of Illinois, Urbana, Ill. (1965).

Peirson, D. H., Cambray, R. S., Spicer, G. S., Lead-210 and polonium-210 in the atmosphere, Tellus *18*, 427-433 (1966).

Peters, B., Cosmic ray produced radioactive isotopes as tracers for studying large-scale atmospheric circulation. J. Atm. Terr. Phys. *13*, 351-370 (1959).

Pollak, L. W., Murphy, T., Sampling of condensation nuclei by means of a mobile photoelectric-counter, Archiv f. Meteorol., Geophys. u. Bioklim. (A) *5*, 100-119 (1952).

Reiter, R., Reiter, M., Relations between the contents of nitrate and nitrate ions in precipitations and simultaneous atmospheric electric processes. IN: "Recent Advances in Atmospheric Electricity" (Edited by L. G. Smith), 175-194, Pergamon Press, New York 1958.

Reiter, R., Carnutz, W., and Sladkovic, R., Effects of atmospheric fine structure characteristics on the vertical distribution of aerosols, Arch. Met., Geoph. Bioklim. (A) *17*, 336-365 (1968).

Revelle, R., Suess, H. E., Carbon dioxide exchange between atmosphere and ocean, and the question of an increase of atmospheric CO_2 during the past decades, Tellus *9*, 18-27 (1957).

Riezler, W., Walcher, W., Kerntechnik, Stuttgart 1958.

Robbings, R. C., Cadle, R. D., and Eckhardt, D. L., The conversion of sodium chloride in the atmosphere, Journ. of Meteorol. 16, 53-56 (1959).

Rossano Jr., A. T., Air pollution control-Guidebook for management, Environmental Science Division, E.R.A. Inc., Stanford, Connecticut 1969.

Runge, H., Blaue Sonne-blauer Mond, (Blue sun-blue moon), Zeitschr. f. Meteorol. *5*, 60-62 (1951).

Sagalyn, R. C., Faucher, G. A., Aircraft investigations of the large ion content and conductivity of the atmosphere and their relation to meteorological factors, Journ. Atm. Terr. Physics *5*, 253-273 (1954).

Schmauss, A., Wigand, A., "Die Atmosphäre als Kolloid," (The atmosphere as a colloid), Sammlung Vieweg, No. 96, Verlag Vieweg, 74 (1929).

Schulz, L., Beiträge zur kenntnis der Luftionen, (Contribution toward a knowledge of air ions), ZS. f. d. ges. Physik, Therapie *45,* 120-144 (1933).

Schumann, G., Ursprung und Verbreitung des radioaktiven Fallouts, (Origin and distribution of radioactive fallout), Nat. Wiss. 54, 6 (1967).

Seiler, W., Junge, C., Decrease of carbon monoxide mixing ratio above the polar tropopause, Tellus *21*, 447-449 (1970).

Stergis, C. G., Study of atmospheric ions in a nonequilibrium system, Geophys. Res. Papers AFCRC (Air Force Cambridge Research Center) No. 26, Cambridge, Mass. 1954.

Stern, A. C., Air Pollution, Vol. I, *25*, Academic Press, New York 1962.

Stern, A. C., Air Pollution (second Edition), Vol. I: Air Pollution and its Effects (p. 694), Vol. II: Analysis, Monitoring and Surveying (p. 684), Vol. III: Sources of Air Pollution and their Control, Academic Press, New York 1968.

Strom, G. H., Atmospheric dispersion of stack effluents, in: Air Pollution Vol. I (A. C. Stern, Ed.), Academic Press, New York 1968.

Suess, H. E., Bull Atm. Scientist *17*, 374 (1961).

Suess, H. E., Radiocarbon concentration in modern wood, Science *122*, 415-417 (1955).

Tebbens, B. D., Gaseous pollutants in the air, in: Air Pollution Vol. I (A. C. Stern, Ed.), Academic Press, New York 1968.

Thomson, W., (Lord Kelvin), On the equilibrium of vapour at a curved surface of a liquid, Philos. Mag., Ser. 4, *42*, 448-452 (1871).

Tow, P. S., Journal Air Poll. Control Assoc. 7, 234-240 (1957).

Twomey, S., Severynse, G. T., Journ. Atmosph. Science *21*, 558 (1964).

U.S. Dept. Health, Education and Welfare, Air Quality Criteria for Particulate Matter, NAPCA Publication No. AP-49 (1969).
US Dept. HEW, Nationwide inventory of air pollutant emissions. APCA-Publication No. AP-73 (1968).

Verzar, F., Höhenklima-Forschungen IV: Atmospharische Konden-sationskerne, (Altitude climate research IV: Atmospheric con-densation nuclei), Basel/Stuttgart (1967).
Volz, F., Optik des Dunstes. In: Handbuch der Geophysik, (Optics of haze), (Editors: F. Linke and F. Möller), vol. VIII, 823-897 (1956).
Volz, F., Scattering in the atmosphere during twilight and ozone absorption, Symposium on Radiation, Wien 1961.

Weger, N., Luftkörper und Größenverteilung atmosphärischer Ionen, (Air masses and size distribution of atmospheric ions), Gerl. Beitr. z. Geophys. *42*, 331-350 (1934).
Weickmann, H., Recent measurement of the vertical distribution of Aitken nuclei. In: "Artificial Stimulation of Rain," (Proc. 1st Conf. Phys. Clouds and Precipitat. Particles) Pergamon Press, New York, 81-88 (1957).
Whitby, K. T., Clark, W. E., Electric aerosol particle counting and size distribution measuring system for the 0.015 to 1μ size range, Tellus *18*, 573-586 (1966).
Wieland, W., Eine neue Methode der Kondensationskernzählung, (A new method for counting condensation nuclei), Eidgen. Komm. z. Studium der Hagelbildung und -abwehr, Wissensch. Mitteilungen No. 6, E.T.H. Zürich 1955.
Wilson, R., The blue sun of 1950 September, Monthly Notices Royal Astron. Soc. *3*, No. 5, 478-489 (1951).
Wilson, J. T., Russel, R. D., Farquhar, R. M., Radioactivity and age of minerals, Handbuch der Physik (S. Flugge), 47. Band, 288-363 (1956).
Woodcock, A. H., Salt nuclei in maritime air as a function of alti-tude and wind force, Journ. of Meteorol. *10*, 362-371 (1953).
Wright, H. L., The size of atmospheric nuclei: Some deductions from measurements of the number of charged and uncharged nuclei at Kew observatory, Proc. Phys. Soc. 48, 675-689 (1936).

Zebel, G., Zur Theorie der Koagulation elektrisch ungeladener Aerosole, (On the theory of coagulation of electrically neutral aerosols), Kolloid-Zeitschrift *156*, 102-107 (1958).

Zebel, G., Coagulation of aerosols. In: "Aerosol Science," (edited by C. N. Davies), Academ. Press, 31-58 (1366).

Index